J.A. FOWLER

Guide to the Reptiles and Amphibians of the Savannah River Site

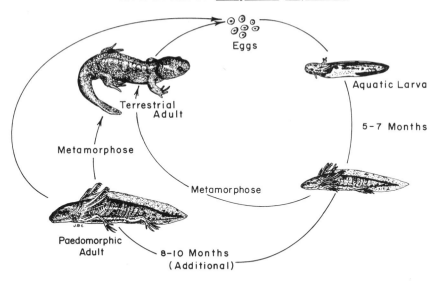

Guide to the Reptiles and Amphibians of the Savannah River Site

J. Whitfield Gibbons
Raymond D. Semlitsch

The University of Georgia Press
Athens and London

© 1991 by the University of Georgia Press
Athens, Georgia   30602
All rights reserved

The paper in this book meets the guidelines for permanence
and durability of the Committee on Production Guidelines
for Book Longevity of the Council on Library Resources.

Printed in the United States of America

95 94 93 92 91   5 4 3 2 1

Library of Congress Cataloging in Publication Data

Gibbons, J. Whitfield, date.
    Guide to the reptiles and amphibians of the
Savannah River Site / J. Whitfield Gibbons, Raymond
D. Semlitsch.
        p.   cm.
    Includes bibliographical references and index.
    ISBN 0-8203-1277-0 (alk. paper)
    1. Reptiles—South Carolina—Aiken Region.
 2. Amphibians—South Carolina—Aiken Region.
 3. Reptiles—Savannah River Region (Ga. and S.C.).
 4. Amphibians—Savannah River Region (Ga. and
S.C.).   I. Semlitsch, Raymond D.   II. Title.
QL653.S6G53   1991
597.6'09757'75—dc20   90-41584
                                                CIP

British Library Cataloging in Publication Data available

This book is dedicated to those who have devoted their time, ideas, and enthusiasm to the study of reptiles and amphibians at the Savannah River Ecology Laboratory.

Contents

List of Illustrations	ix
Acknowledgments	xi
Introduction	1
Habitats, Collecting Methods, and Taxonomy	3
Key to the Orders of Amphibians and Reptiles	26
Key to the Salamanders	27
Key to the Frogs and Toads	38
Key to the Turtles	49
Key to the Lizards	52
Key to the Snakes	54
Species Accounts	61
Salamanders	61
Frogs and Toads	68
Alligators	75
Turtles	75
Lizards	81
Snakes	83

viii *Contents*

Problem Species 94

    Unresolved Records 94
    Fall Line Subspeciation 97
    Venomous Species 97
    Edible Species 97
    Introduced Species 98
    Endangered Species 98
    Herpetological Research Problems 98

Bibliography 101

    Subject Categories for SREL Herpetological Reprints 101
    Species of Reptiles and Amphibians for Bibliography 105
    SREL Herpetological Reprints 110
    Theses and Dissertations Based on Herpetological Studies
      on the SRS 124
    Additional Useful References 125

Species Index 129

Illustrations

Maps

Map 1. The state of South Carolina, showing the location of the Savannah River Site / *4*

Map 2. The state of South Carolina divided into physiographic regions / *5*

Map 3. General map of the Savannah River Site / *7*

Map 4. Herpetological collecting sites on the Savannah River Site / *10*

Plates

Plate 1. Larval *Siren intermedia* and *S. lacertina* / *30*

Plate 2. Larval *Eurycea quadridigitata* / *31*

Plate 3. Larval *Necturus punctatus* / *31*

Plate 4. Larval *Desmognathus auriculatus* / *32*

Plate 5. Larval *Notophthalmus viridescens* / *33*

Plate 6. Larval *Ambystoma talpoideum, A. opacum, A. cingulatum, A. mabeei, A. tigrinum, A. maculatum* / *34*

Plate 7. Ventral view of larval *Ambystoma talpoideum, A. opacum, A. tigrinum* / *35*

x *Illustrations*

Plate 8. Larval *Pseudotriton ruber* and *P. montanus* / 36

Plate 9. Larval *Eurycea cirrigera* and *E. longicauda* / 36

Figures

Figure 1. Long-term monthly mean temperatures and precipitation for the Savannah River Site / 6

Figure 2. Salamander costal grooves and nasolabial groove / 27

Figure 3. Frontal view of tadpole mouthparts / 42

Figure 4. Body size measurements of tadpole / 42

Figure 5. Positions of tadpole anal tube and eyes / 43

Figure 6. Plastrons of *Pseudemys concinna* or *P. floridana* and *Trachemys scripta* / 50

Figure 7. Carapace shapes of *Trachemys scripta* and *Deirochelys reticularia* / 50

Figure 8. Tail scale patterns in *Eumeces* / 52

Figure 9. Labial scale pattern in *E. laticeps* / 53

Figure 10. Tail scale patterns in crotalid and colubrid snakes / 55

Figure 11. Head of *Crotalus*, showing numerous small scales / 55

Figure 12. Solid black tail of *Crotalus horridus* / 55

Figure 13. Counting system for dorsal scale rows / 55

Figure 14. Comparison of keeled and smooth scales / 56

Figure 15. Lateral line scale patterns of *Thamnophis sauritus* and *T. sirtalis* / 58

Figure 16. Upturned rostral scale of *Heterodon simus* / 59

Figure 17. V-shaped mark on head of *Elaphe guttata* / 60

Acknowledgments

Manuscript preparation and much of the research on which the data are based were carried out under Contract EY-76-C-09-0819 between the University of Georgia and the U.S. Department of Energy. Needless to say, the collection of such large numbers of animals requires the cooperation and assistance of numerous individuals. We greatly appreciate the effort extended by so many of our colleagues during the past several years.

Particular thanks are given to Karen K. Patterson and Judith L. Greene, who were instrumental in compiling herpetological information of the SRS through 1977. The original report by Gibbons and Patterson ("Reptiles and Amphibians of the Savannah River Plant," 1978) served as the foundation for this book.

Jean Coleman made all of the drawings except those of the larval salamanders, which were done by Molly Griffin. Manuscript preparation was aided by Cathy Houck, Miriam Stapleton, Marianne Reneau, Loretta Bracco, Sarah Collie, and Marie Fulmer.

Chuck Segal did an outstanding job of formatting and organizing the manuscript for final publication. Judy Greene and Marie Fulmer were essential in the onerous tasks of compiling records and editing copy.

This guide has profited greatly from the expertise and advice of a large number of herpetologists who have been at SREL as staff, students, or visiting investigators. Direct contributions in the compilation of this report were made by particular individuals in the following manner: Bill Cooper provided species descriptions for the three species of *Eumeces*; Judy Greene provided species descriptions for the turtles; Grady Knight provided information on *Siren* and *Amphiuma*; Rich Seigel and Ray Loraine provided a substantial amount of information on *Seminatrix*; Larry Wilson provided observational field notes based on several collecting trips; Jim Knight reviewed several sections and provided species descriptions for certain snakes; Trip Lamb provided species descriptions for frogs; Nancy Martin provided information for habitat descriptions; Sarah Oliver compiled much of the snake data; Marie Fulmer proofread the manuscript numerous times. Rich Seigel, Janalee Caldwell, and Laurie Vitt provided critical reviews of the manuscript during early drafts. We particularly thank Jeff Lovich for his help in consolidating material and editing during the final stages of manuscript preparation. Jim Knight and Judy Greene compiled much of the cross-referenced bibliography and offered helpful comments on the manuscript. Susan Novak edited the final manuscript and dealt with the tedium of cross-checking references. We also are grateful to I. Lehr Brisbin, Carl Strojan, and Janell Gregory for the time devoted to review of the final manuscript and for general support from the SREL National Environmental Research Park (NERP) program.

We especially thank Marie Fulmer and Chuck Segal for their efforts in computer transfers and compilation during the final proofing and consolidation of materials. Patricia J. West did an outstanding job of editing the final copy of the manuscript.

Our sincere thanks go to Win Seyle (Savannah Science Museum), Julian Harrison (College of Charleston), and Steve Bennett (South Carolina Wildlife and Marine Resources Department) for their reviews of the original manuscript and their insightful comments about regional herpetofauna. Fred Turner made numerous recommendations that greatly improved an earlier version. Special thanks go to Robert Mount (Auburn University), whose thorough and herpetologically insightful editorial comments in an anonymous review resulted in the authors' request for the name of the reviewer from the University of Georgia Press.

Introduction

The objective of this guide is to provide taxonomic, distribution, and ecological information about the reptiles and amphibians of the U.S. Department of Energy's Savannah River Site (SRS) located on the Atlantic Coastal Plain south of Aiken, South Carolina. The foundation of herpetological information is based on general and regional publications and on extensive research by local and visiting investigators. Much of the site-specific research has been published in the scientific literature or is available in theses, dissertations, and reports. Our guide attempts to consolidate the findings from all these efforts in order to present, as cohesively as possible, our present understanding of the ecology and distribution of species of reptiles and amphibians that occur locally. The intent is to provide sufficient comprehensive information to other biologists, regardless of their experience in herpetology, to permit them to undertake studies that in some manner incorporate the herpetofauna of the SRS.

The guide should also be useful to any wildlife observer, whether amateur or professional, as a comprehensive overview of ecological information about most of the reptiles and amphibians of South Carolina and Georgia.

No account of the herpetofauna of South Carolina has been published. Checklists of coastal species, however, have been presented (Gibbons, 1978, reptiles; Harrison, 1978, amphibians). General herpetofaunal accounts applicable to South Carolina have been published in accounts of all eastern reptiles and amphibians (Cochran and Goin, 1970; Conant, 1975; Behler and King, 1979; Smith and Brodie, 1982), and in specific accounts of U.S. turtles (Carr, 1952; Ernst and Barbour, 1972), snakes (Schmidt and Davis, 1941; Wright and Wright, 1957), lizards (H. M. Smith, 1946), alligators (Neill, 1971), salamanders (Bishop, 1947), and frogs and toads (Wright and Wright, 1949). The most pertinent regional works are *Reptiles and Amphibians of Alabama* (Mount, 1975) and *Amphibians and Reptiles of the Carolinas and Virginia* (Martof et al., 1980). Martof's *Amphibians and Reptiles of Georgia, A Guide* (1956) is outdated in many respects and consequently is of limited value. Ashton and Ashton (1981, 1985) provide general information on the reptiles of Florida. The most thorough accounts of the status of SRS herpetology are presented by Gibbons et al. (1976) and Gibbons and Patterson (1978).

Each species that has been verified to occur on the SRS on the basis of voucher specimens, or that has been reported to occur, is considered. Most voucher specimens are catalogued as part of the Savannah River Ecology Lab (SREL) Museum collection, although particular specimens may have been donated to and housed at other museums. Most species are probably represented by a single subspecies on the SRS. However, subspecific assignments are not given in the present guide because few taxonomic studies have been undertaken on the SRS to confirm subspecific status. The scientific names used are those we considered to be most current on the basis of recent publications or discussions with colleagues. The mercurial nature of scientific taxonomy makes it highly unlikely that we will have chosen the name in every instance that will still be in use a few years from now. The information in each species account pertains to our experience with the species on the SRS. Detailed ecological information about many of the species can be obtained directly from the SREL references given at the end of each account. This guide should be considered as a general overview of the herpetofauna of the SRS that will serve as a starting point for anyone interested in a particular group or species. The systematics, ecology, and life history of the SRS herpetofauna described here should be a foundation to which information can, and surely will, be added.

Habitats, Collecting Methods, and Taxonomy

HABITATS

Regional Aspects

The SRS is located in the west-central portion of South Carolina and encompasses portions of Aiken, Barnwell, and Allendale counties (map 1). The site's southwest boundary is the Savannah River, a typical large southern river with extensive floodplains and oxbow lakes. The northern boundary is approximately 32–48 km south of the Fall Line. Throughout a major portion of the Southeast, the Fall Line represents the transitional zone between montane or piedmont and coastal plain environments. It also forms a northern boundary for numerous species found on the SRS and is a zone of intergradation for many others. The entire SRS site lies within the Atlantic Coastal Plain Physiographic Province (map 2). The SRS acreage includes several major soil types, most composed of sand overlying sandy clay loam.

The summer climate of South Carolina is generally hot and humid, and winters are usually mild (figure 1). Normal January temperatures for the upper coastal plain region, including the SRS, are lows near freezing and highs around 13°C. Normal July temperatures are lows around 21°C and highs around 32°C. Normal yearly precipitation is 91–112 cm, primarily as rainfall. Snow is uncommon. Thirty-year average meteorological data for the SRS are shown in figure 1.

Local Aspects

The SRS encompasses approximately 780 km^2. A major portion of the tract is protected from public intrusion and has the typical array of habitats

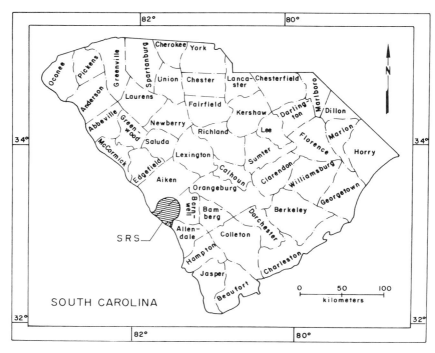

Map 1. The state of South Carolina, showing the location of the Savannah River Site in Aiken, Barnwell, and Allendale counties within the latitude and longitude coordinates 33°05' N–33°25' N, 81°30' W–81°50' W.

characterizing nonurban, nonagrarian portions of the upper coastal plain of South Carolina. The SRS has five nuclear production reactors, of which three are currently operable, and vast quantities of water are used for cooling purposes. These waters have been, and in some instances still are, released into a variety of aquatic habitats including reservoirs, thermal canals, streams, stream deltas, and swamps.

Predominant freshwater habitats include the Savannah River and five tributary freshwater streams, a 1700-ha reservoir system, numerous Carolina bays, and a few abandoned farm ponds and minor impoundments. The cypress-gum swamps and lowland hardwood forests bordering the river and its tributaries comprise 10–15% of the site. Pine plantations and natural pine stands make up about 40% of the area.

Several upland hardwood stands are scattered throughout the SRS but constitute less than 4% of the site. The remainder of the SRS is composed of mixed hardwood and pine, aquatic and semiaquatic habitats, abandoned

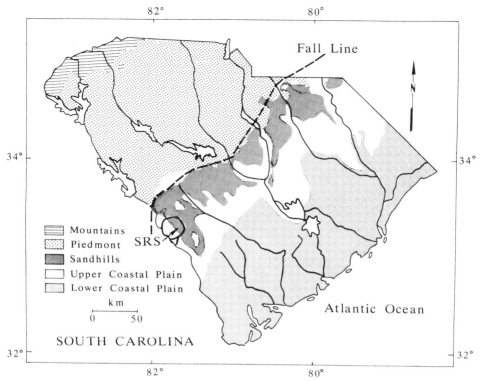

Map 2. The state of South Carolina divided into physiographic regions. The location of the Fall Line is indicated by the dashed line running northeast to southwest across the center of the state.

old fields, industrial complexes, and an extensive highway system (map 3). Many of the natural and affected habitats have been identified as research set-aside areas (Hillestad and Bennett, 1982; map 3).

The SRS was acquired from private lands in 1951 by the U.S. government. At the time of acquisition, 30–40% of the area was farmed (primarily cotton and corn) and the remainder was mostly second-growth pine or hardwood forests. During the 38 years since the site was established, the most extensive environmental impact has resulted from U.S. Forest Service forest management programs. Most of the abandoned farmland has been planted in pine or is undergoing natural succession toward turkey oak–longleaf pine associations, an edaphic climax community in this region.

Extensive draining has not been done on the site since its establishment, and most lowland areas have remained undisturbed for more than a third

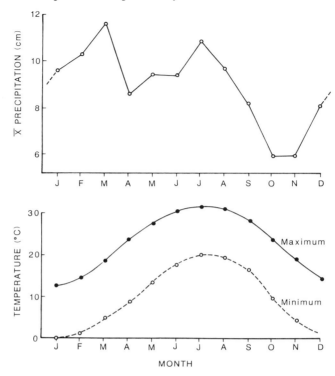

Figure 1. Long-term monthly mean temperatures and precipitation for the Savannah River Site, South Carolina, based on 1955–1986 records from Bush Field, approximately 30 km from Augusta, Georgia.

of a century. Major aquatic alterations have resulted from thermal releases into three of the five tributary streams and from construction of the Par Pond reservoir system in 1958 (Gibbons and Sharitz, 1974, 1981) and the L-Lake reservoir in 1985. Selected habitats deserve specific mention because of their uniqueness to the site or region and their importance to the herpetofauna.

Aquatic Habitats

Carolina Bays

These unique aquatic environments, occurring primarily across the coastal plain regions of Georgia and the Carolinas and as far north as Maryland,

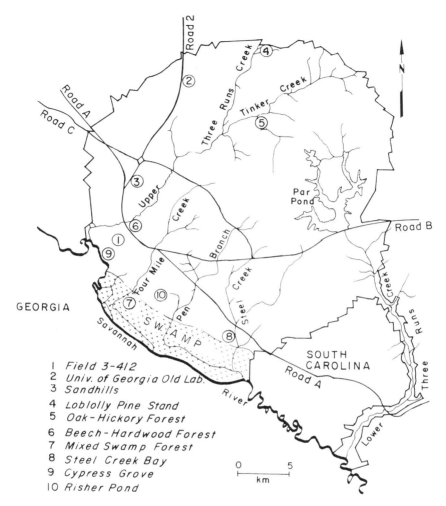

Map 3. General map of the Savannah River Site, indicating National Environmental Research Park (NERP) set-aside areas (Hillestad and Bennett, 1982) and the Savannah River swamp.

are the primary freshwater lentic habitats occurring naturally on the SRS (Schalles, 1979; Sharitz and Gibbons, 1982; Schalles et al., 1989). At least 200 are present on the site. The geologic origin of the Carolina bays is unknown, but they are characteristically ovate in shape, oriented in a northwest–southeast direction, and have seasonally or annually fluctuating water levels. Most bays on the SRS are temporary, filling with rainwater in the winter and drying each summer, although a few contain water throughout the year during most years. Those on the SRS have no tributary water supply so water levels depend on precipitation and evaporation. Some Carolina bays support populations of fish. These habitats are extremely productive as breeding or feeding sites for some species of herpetofauna and have been the focal point of many SRS studies. The following Carolina bays have been used extensively in herpetofaunal research.

ELLENTON BAY
This bay is a natural freshwater habitat and a typical Carolina bay. The basin, representing the high-water level of Ellenton Bay, covers approximately 10 ha. A road embankment 5–6 m wide divides the bay and creates two completely separate aquatic areas. Water surface area and depth are extremely variable. Maximum depth when the basin is full is more than 2 m. In 1955–56 and again during the summers of 1968, 1981, 1985, 1986, and 1987 Ellenton Bay dried almost completely, with water remaining only in three or four small pools a few centimeters deep and 2–3 m across. In 1986, no standing water remained. Even during these periods of drought, however, standing water of more than 0.5 m depth normally remains beneath the thick organic crust that covers the entire lake basin. Many areas of the basin remain quite mucky, so that during dry spells as much as a hectare of viscous mud surrounds the area of open water. Ellenton Bay is located near one edge of Field 3-412, which was abandoned in 1952 and has been the site of many previous ecological studies (Odum and Kuenzler, 1963; Golley and Gentry, 1966; Van Pelt, 1966). Predominant plants peripheral to the basin are dog fennel (*Eupatorium* sp.), lespedeza (*Lespedeza* sp.), and blackberry (*Rubus* sp.). The white water lily (*Nymphaea odorata*) and water shield (*Brasenia* sp.) are the most evident aquatic plants.

RAINBOW BAY
This Carolina bay is approximately 1 ha in area, has a maximum water depth of about 1 m, and dries each year during the summer (map 4). Herbaceous plants common to the basin are knotgrass (*Scirpus cyperinus*), creeping rush (*Juncus repens*), common cattail (*Typha latifolia*), and spike-rush (*Eleocharis* sp.). Buttonbush (*Cephalanthus occidentalis*) and cypress (*Taxodium* sp.) are also found in the basin. The aquatic area is surrounded by deep, well-drained sandy soil vegetated with slash pine (*Pinus elliottii*)

and loblolly pine (*P. taeda*) plantations of various ages with hardwoods along the water's edge, including sweet gum (*Liquidambar styraciflua*), water gum (*Nyssa sylvatica*, var. *biflora*), water oak (*Quercus nigra*), and wax myrtle (*Myrica cerifera*). Thick vegetation consisting of blackberry (*Rubus* sp.), honeysuckle (*Lonicera* sp.), and greenbriar (*Smilax rotundifolia*) separates the aquatic area from the surrounding upland pine plantations.

SUN BAY

This Carolina bay is approximately 1 ha in area, has a maximum water depth of 0.35 m, and dries each summer. This site was altered in June 1978 when a ditch was dug to partially drain it in preparation for construction. As a consequence this bay contained about 25% of the water volume of Rainbow Bay and dried earlier each year. During 1982 and 1983, construction continued to the point where Sun Bay now no longer contains water. Herbaceous plants in the basin of Sun Bay were similar to those of Rainbow Bay, although cypress trees were absent. The upland habitat around Sun Bay is dominated by loblolly pine and longleaf pine (*P. palustris*) plantations with pockets of mixed sweet gum and oaks (*Quercus* spp.). Approximately 30% of the perimeter of Sun Bay was bordered by a 20-ha clear-cut field planted in loblolly pine in 1976.

FLAMINGO BAY

This Carolina bay is approximately 5 ha in area and has a maximum water depth of 1 m. Although Flamingo Bay has dried three times for brief periods in the last 20 years (1968, 1981, 1986), the habitat probably functions as permanently aquatic to most of the amphibians and reptiles using it. The surrounding upland habitats are slash and loblolly pine plantations. The perimeter of the aquatic habitat is dominated by sweet gum, oaks, red maple (*Acer rubrum*), and wax myrtle. The basin contains buttonbush, water gum, bulrush, spikerush, lizard's tail (*Saururus cernuus*), and yellow water lilies (*Nelumbo lutea*).

MORSE CODE BAY

This Carolina bay is approximately 2 ha in area, has a maximum water depth of 0.5 m, and usually dries each summer. The bay is open, and vegetation such as *Scirpus*, *Juncus*, and *Cephalanthus* covers the basin. It is surrounded by pine plantations interspersed with pockets of mixed hardwoods.

CAROLYN'S BAY

This bay is adjacent to Morse Code Bay (within 50 m at high water) and very similar in area and depth. The basin of Carolyn's Bay is vegetated

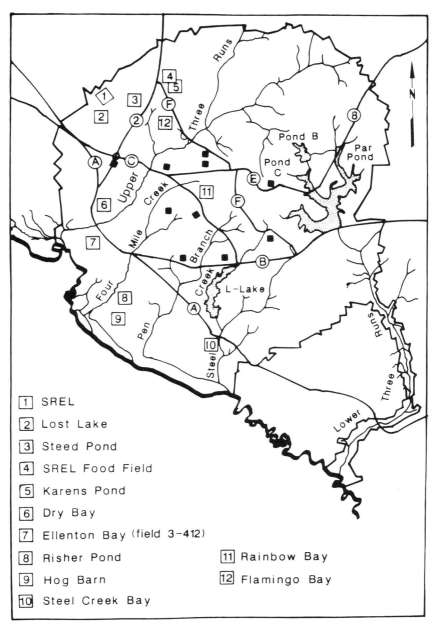

Map 4. The Savannah River Site, with selected herpetological collecting sites identified. Reactor facilities are indicated by black squares.

almost entirely by grasses or sedges and surrounded by pine plantations and pockets of mixed hardwoods.

DRY BAY
This Carolina bay contains water year-round in almost all years. Fish are normally present. The only recorded times of almost total drying were during the autumns of 1968 (when it was first discovered by SREL and paradoxically named "Dry Bay"), 1981, and 1986. A thorough description is given by Sharitz and Gibbons (1982).

WOODS BAY
This small bay is approximately 200 m south of Ellenton Bay. It is surrounded by hardwoods and pines and dries completely every year, but when water is present, it is known to harbor some turtles from the Ellenton Bay populations.

STEEL CREEK BAY
This is a large, open aquatic habitat approximately 5 ha in area. During the drought of 1981 the only water remaining was in a "gator hole" at the north end of the lake where the water was still about 1 m deep and several fish species were present.

LOST LAKE
Lost Lake was a 12-ha Carolina bay that received effluent from an industrial facility and supported no emergent or submerged aquatic macrophytes, due to heavy concentrations of herbicides and major fluctuations in water level. Despite the lack of macrophytes, the lake supported a large population of turtles and served as a breeding site for several species of amphibians. In 1989 the basin of the bay was bulldozed during a project designed to prevent radioactively contaminated sediments from dispersing. The former lake basin is surrounded primarily by stands of slash and loblolly pine.

Par Pond and L-Lake Reservoir Systems

An extensive man-made reservoir habitat exists on the SRS. The Par Pond reservoir system is primarily a closed loop of man-made canals and lakes, including Pond C (67 ha), Pond B (81 ha), and Par Pond (1100 ha). Pond C temperatures may exceed 50°C during periods of reactor operation, although in some areas of the lake the temperatures remain substantially lower. Water at temperatures exceeding 35°C enters Par Pond from Pond C at the "hot dam" and disperses throughout the hot arm of the reservoir. The remainder of Par Pond (north arm and west arm) is at temperatures normal or only slightly above normal for the area. Pond B and the north arm of Par Pond received thermal effluent from 1958 to 1964 and are now in a post-thermal state.

L-Lake is a 405-ha reservoir that was formed in 1985 by damming the upper one-third of Steel Creek below L Reactor. The lake is 7000 m long and up to 1200 m wide (Wein et al., 1987). Cooling water from the reactor enters the lake at high temperatures that would be lethal for most organisms, and temperatures are elevated throughout all portions of the reservoir during periods of reactor operation.

Dammed Artificial Ponds

Ponds 2.5–7.4 ha in area are characteristic of the region for use as stocked, warm-water fishing lakes. Several such habitats were left over on the SRS from private ownership and have been focal points for research on many aquatic and semiaquatic species of herpetofauna.

RISHER POND

This small farm pond was constructed during the 1930s by damming a small stream to form a 1.1-ha reservoir with a maximum depth of 2.5 m. Habitats peripheral to the lake are loblolly and longleaf pine plantations on three sides and a lowland swamp deciduous forest on the west side below the dam. Emergent plant species around the lake include cattails (*Typha latifolia*) and rushes (*Juncus* sp.). Water levels do not vary more than a few centimeters seasonally or annually. Risher Pond was drained during the summer of 1984 in an attempt to capture turtles but was refilled within one month.

STEED POND

This pond (Freeman, 1960) was formed by damming a headwater stream of Upper Three Runs Creek. At that time the mean depth was 1.5 m. Willows (*Salix nigra*) and alders (*Alnus serrulata*) grow in the major body of the pond and in surrounding low areas that flood seasonally. A mixed pine-hardwood forest surrounds the pond. The habitat has had a history of raised and lowered water levels. The dam was broken sometime before 1967 and the pond surface was less than 1 ha until 1972. The maximum aquatic area with the dam in place is greater than 4 ha. The dam was removed again during autumn 1985, and since that time the habitat has been mostly dry with the exception of the small stream (Tim's Branch). Runoff waters from Lost Lake may have resulted in contamination of the sediments with low-level radioactive materials (primarily uranium).

LODGE LAKE

This is a shallow flooded area in the tributary of Tim's Branch above Steed Pond. The area was apparently wooded at one time but became flooded due to either beaver activity or road construction. Most of the trees have fallen so that it is now a log-filled, shallow lake with abundant herpetofauna.

FIRE POND

This abandoned farm pond from pre-SRS days harbored the typical herpetofaunal associations expected of such a habitat. It was drained in 1983 and now consists of a marshy area through which a small stream flows.

DICK'S POND

This is a pre-SRS farm pond, approximately 1 ha in area with a mean depth of 1.4 m and surrounded by hardwoods and pines. *Potamogeton diversifolius* and *Sphagnum* sp. are the major submerged plants. Radioactive materials were introduced into the pond for an experiment during the 1970s but the isotopes had short half-lives and the area is not contaminated at this time.

TWIN LAKES

These two lakes (Gus's Pond [lower lake] and Debby's Pond) are each about 1–1.5 ha and average 1–2 m in depth, and are separated by a dam. *Eleocharis acicularis* is a major submerged vascular plant species. The area is surrounded by hardwoods and pines.

Borrow Pits and Seepage Basins

A variety of man-made aquatic habitats are present on the SRS as a result of construction activities. Most have standing water during wet periods and dry by late summer and during prolonged dry periods. Some have been used in herpetological studies.

PICKEREL POND

This borrow pit is part of the Rainbow Bay System along Road C-5. Pickerel Pond is small (0.1 ha in area), has a maximum water depth of 1.2 m, and usually dries each summer. The pond is near Crystal Creek and during wet years may be flooded and invaded by fish from the adjacent stream (e.g., 1979). The basin of Pickerel Pond has little vegetation and the canopy is closed over by sweet gum, red maple, and water gum.

LINDA'S POND

This pond is located on the east side of Road C-5 in the Rainbow Bay System. It probably originated as a borrow pit and developed as an aquatic site when natural drainage was blocked by the construction of the roadbed for Road C-5. It is about 0.5 ha in area, has a maximum water depth of 0.8 m, and dries each summer. Mixed hardwoods (red maple, water gum, sweet gum, and oaks) are found in the basin and produce a closed canopy over the pond.

BULLDOG BAY

This bay (not a Carolina bay) is a small stream system 600 m north of Ellenton Bay that is frequently flooded due to beaver dam-building activities.

BULLFROG POND

This pond is located in the Rainbow Bay System along Road C-5. It probably has the same origin as Linda's Pond; that is, a borrow pit to which natural drainage has been blocked by the roadbed of C-5. Bullfrog Pond is about 0.5 ha in area, has a maximum water depth of 0.58 m, and dries each summer. One-half of the pond is open with *Scirpus* and *Cephalanthus* covering the basin, whereas the other half has a closed canopy of *Liquidambar*, *Acer*, and *Nyssa*.

RISHER ROAD POND

This pond was formed when a small stream was blocked by Risher Road. It is located on the north side of Risher Road about 2–3 km west of Risher Pond. It floods during the winter or early spring, forming a pond in the thick deciduous forest bottomland, and dries to a trickle in summer.

GREEN LAKE

This was an artificial overflow basin between M-Area and Lost Lake. Only two species of reptiles have been noted from the lake: slider turtles (*Trachemys scripta*) that migrated to the habitat during a period of drought at Lost Lake, and a single American alligator of unknown origin. Because of the variety of chemicals, detergents, and other industrial runoff materials, this lake was not considered an ideal habitat for herpetological investigations. Green Lake was filled with dirt during the remediation efforts at Lost Lake in 1989 and no longer exists.

ELLENTON BAY BORROW PIT

This pond is located in a system of small pools created by excavation of sand and gravel near River Road. It is very temporary in nature, often drying by late spring. The pond is about 0.2 ha in area, with a maximum water depth of 0.5 m, and is sparsely vegetated with grasses but has no trees in the basin or along its edge. It is surrounded by pine plantations and deciduous forest.

A-AREA SEEPAGE BASINS

These four basins are located 100 m north of the SREL building. They were constructed in the 1950s for containment of chemical waste materials, including radioactive isotopes. Although barren of emergent vegetation, they have high productivity and the typical herpetofaunal assemblages that would be expected in such habitats. Unfortunately, these basins have appreciable levels of radioactive strontium and cesium in addition to a variety of other isotopes. Their primary use from a herpetological research standpoint is for experimental studies using radioactively contaminated habitats and organisms.

KAREN'S POND

This temporary pond (borrow pit) is approximately 800 m^2 when full, although it dries periodically. Maximum depth is normally less than 1 m. This pond does not dry in years with high rainfall (e.g., 1979). The basin of the pond has little vegetation except for a thick mat of grass that covers the bottom and banks. The pond is encircled by mixed pine and deciduous hardwoods extending several miles to the south, east, and west. Approximately 75 m to the south is a shallow, ephemeral Carolina bay. The area north of the pond is primarily old-field habitat. Buildings and parking lots of the Savannah River Forest Station are to the northwest.

Streams

Pristine blackwater streams were once characteristic of the coastal plain of South Carolina. Today most of the streams carry industrial, domestic, or agricultural pollutants. Upper Three Runs Creek has many of its headwaters on the site and is minimally affected by agricultural runoff or domestic sewage inputs. Parts of this major tributary to the Savannah River originate in upland hardwood habitats on the SRS and in areas a few miles north, and the creek traverses about 20 miles (32 km) of mostly undisturbed terrestrial habitat, including a mile (1.6 km) or more of cypress-gum swamp adjacent to the river. A few industrial inputs (including tritium releases) from SRS activities occur along the lower two-thirds of the stream but have no apparent impacts on the natural communities.

CRYSTAL CREEK

This small stream originates in an area between Rainbow Bay and Road 6 and flows about 2–3 km into Four Mile Creek. Crystal Creek is a typical mud-bottomed, sluggish coastal plain stream in deciduous forest bottomland.

MCQUEEN BRANCH (SANDY CREEK)
This creek originates in the Defense Waste Processing Facility (DWPF) site near Road F and Road 4 and flows 3–4 km into Upper Three Runs Creek. McQueen Branch is sandy bottomed and sluggish, and flows through deciduous forest bottomland.

PEN BRANCH AND FOUR MILE CREEK
Pen Branch and Four Mile Creek, respectively, receive effluents from C and K reactors and are too hot thermally for reptiles and amphibians throughout much of their main courses when the reactors are in operation. However, certain species can be found along corridors and backwater seepage areas that have slightly elevated temperatures. Both enter the main Savannah River swamp, and much of the warm water is dissipated in the swamp before it reaches the Savannah River.

TINKER CREEK
This creek is a tributary that flows about 17 km into Upper Three Runs Creek and passes through primarily hardwood areas. It is a clear, typical blackwater stream receiving little or no pollution of any sort.

BEAVER DAM CREEK
This creek receives effluent from a coal-fired power station at the heavy water plant, but the thermal elevation diminishes as the creek approaches the river swamp and Savannah River.

STEEL CREEK
This 15-km-long stream originates near P reactor and from 1954 to 1968 received thermal effluent from L reactor. During 1984 much of the forested area of the upper reaches above the floodplain delta was cleared for construction of L-Lake, which now covers 405 ha.

LOWER THREE RUNS CREEK
This creek originates as the overflow from Par Pond and was an original stream system on the SRS. This stream transported thermal effluent from 1953 to 1958. Since that time the impoundment of its upper extremity to create Par Pond has reduced stream flow considerably and altered water quality because of input into Par Pond from the Savannah River. It is contaminated by low levels of cesium not considered to be high enough to be of concern by individuals working in the stream. Lower Three Runs is

approximately 22 km in length from the Par Pond cold dam to its entry into the Savannah River and passes through lowland swamp forest over much of its course.

MILL CREEK

This clear, blackwater tributary to Upper Three Runs Creek is approximately 6 km in length. The surrounding terrain is predominantly hardwood forest.

Cypress-Tupelo Swamp and Bottomland Hardwood Forests

Bottomland hardwood forest habitat is the largest natural vegetation assemblage on the SRS (13,327 ha). In this habitat, small variations in elevation strongly influence the tree species composition. In the swamp, where standing water is present much of the year, bald cypress and swamp tupelo are the dominant trees. In more mesic areas sweet gum, red maple, ash, water oak, and yellow poplar are common species. Approximately 4200 ha are cypress-tupelo swamp forests where standing water is frequent. Water levels fluctuate seasonally (highest in winter and spring; lowest in summer and autumn) and may vary more than 2 m over a year. Rapid changes of more than 1 m may occur as a result of heavy rainfalls or adjustments in water levels at Clark's Hill Reservoir, approximately 56 km upstream (Duke, 1984). These areas are the primary habitats for many species of reptiles and amphibians. The most intensively studied swamp habitat has been that bordering, and in the delta of, Steel Creek.

Savannah River

The southwestern border of the SRS is a large river (total basin approximately 28,000 km^2) that directly receives the major tributaries Upper Three Runs Creek, Steel Creek, and Lower Three Runs Creek. According to Bennett and McFarlane (1983), the river has a gradient of 0.12 m/km in the vicinity of the SRS. The mean annual discharge is 316 m^3/s (McFarlane et al., 1978, 1979). Peak flow and maximum variability in flow occur during March and April, while the lowest flow and least variability are found during the summer and autumn. The 7-day, 10-year minimum flow at the SRS site is estimated to be 160 m^3/s. At flood stage the water mass breaches the channel to form a floodplain up to 3 km wide, and flow may approach 1200 m^3/s. Stream velocity is approximately 0.74 m/s at mean annual discharge and 0.65 m/s at 7-day, 10-year low flow. The river's characteristics are currently dominated by the release of hypolimnetic water from Clark's Hill Reservoir. The principal effects have been a decrease in the incidence of both extreme high and low discharges and a decrease in the average river temperature by 3°C year-round.

Terrestrial Habitats

Abandoned Old Fields

Several hundred acres of fields that were previously cultivated have not been disturbed since the development of the site in the 1950s and have thus undergone more than 38 years of secondary succession. Many of these old-field habitats still remain in an unforested, open habitat state. A few species of reptiles and amphibians are characteristically associated with such successional habitats.

FIELD 3-412

This habitat has been extensively studied by SREL ecologists since the 1950s. It was originally an approximately 500-acre cotton and corn field that was allowed to undergo natural succession. Much of this area still remains untouched except for manipulation from experimental studies. The present composition of the vegetation includes primarily old-field species such as *Lespedeza* and *Andropogon*, but with scattered pine trees throughout the area. Ellenton Bay is located at one edge of Field 3-412.

SREL FOOD FIELD

This is a 16-acre habitat from which all trees were removed in 1969 for an experimental study of old-field succession (Briese and Smith, 1974). The drift fences that encircle the field yielded a large number of captures of reptiles and amphibians. Many were presumably residents of adjoining natural habitats or associated with aquatic areas in the vicinity. The food field is located approximately 200 m north of the old SREL facility.

PINE MANAGEMENT AREAS

A major program of pine planting and harvesting by the U.S. Forest Service has resulted in a complex array of pine plantations and open clear-cut habitats. These range from several-acre clear-cut areas to mature pine plantations with limited undergrowth. Herpetofaunal studies have been conducted in these areas using drift fences and pitfall traps, and the species composition and relative abundance of reptiles and amphibians are known to some extent.

Natural Terrestrial Habitats

The natural terrestrial habitats on the SRS are characteristic of the Atlantic Coastal Plain Physiographic Province. No extensive virgin forest occurs on the site. With the exception of 8.3 acres of old-growth bottomland forest known as Boiling Springs Natural Area along Lower Three Runs Creek, all natural areas are presumably second-growth pine, hardwood, or mixed pine-hardwood.

UPLAND FOREST

This general habitat constitutes approximately 1120 ha on the SRS. Oak-hickory hardwoods are present on the more fertile dry uplands. The common species are white oak (*Quercus alba*), southern red oak (*Q. falcata*), post oak (*Q. stellata*), pignut hickory (*Carya glabra*), and mockernut hickory (*C. tomentosa*).

SCRUB OAK–PINE FOREST

This general habitat constitutes approximately 5425 ha on the SRS. The xeric, sandy areas are typically dominated by turkey oak (*Quercus laevis*), longleaf pine (*Pinus palustris*), blackjack oak (*Q. marilandica*), and scrubby post oak (*Q. margaretta*).

Habitats Environmentally Altered by Man

THERMALLY ALTERED AREAS

The five plutonium-production reactors on the SRS have created a variety of unnatural aquatic environments in regard to thermal characteristics and associated habitat changes (e.g., standing dead trees in some areas). A major influence of the reactor effluents on streams and ponds has been an increase in primary and secondary productivity in these and contiguous areas. Two of the reactors were placed on standby several years ago and are not operating. The resulting termination of heated effluent to some aquatic areas has created lentic and lotic post-thermal habitats. Influences of hyperthermal and post-thermal environments on the biology of some reptiles and amphibians have been studied.

HIGHWAYS

More than 200 km of paved highways and 2000 km of unpaved secondary roads occur on the SRS. Most are not heavily traveled except during shift changes, and many have little traffic at any time. Road collecting has been a relatively successful technique for capturing reptiles and amphibians.

Abandoned Homesites and Buildings

GUNSITE 51
Gunsite 51, located on Road A-2, was an abandoned military barracks site that was razed in the 1960s. The present habitat consists of blackberry, honeysuckle, and other terrestrial plants. An artesian well located at this site flows south toward Upper Three Runs Creek. Until autumn 1986, a large amount of broken concrete, bricks, and sheet metal at the site provided cover and hibernation sites for herpetofauna. Snakes and lizards are frequently found in the area.

OLD ELLENTON
This was the site of the original town of Ellenton, which was razed in the 1950s. It is vegetationally aberrant because of the large variety of cultivated shrubs, vines, and trees prevalent in the approximately 1 km^2 that constitutes the abandoned town. Bricks and other rubble can be found in certain sections of this habitat and may provide hiding places for certain reptiles and amphibians.

ASHLEY PLANTATION
Ashley Plantation comprised a group of small buildings and a large house that were razed in the 1950s. For many years numerous board piles provided cover for reptiles and amphibians on the site. Many of them have now been removed.

HOG BARN
This is a large cypress-wood barn located at the margin of the Savannah River swamp between Four Mile and Pen Branch. Because of the setting alongside the swamp, a number of herpetological studies have been conducted in the area. The building itself does not necessarily attract any particular species of reptile or amphibian, except for *Anolis carolinensis*.

COLLECTING METHODS

Most collecting of reptiles and amphibians on the SRS has employed one of three techniques: terrestrial drift fences with pitfall traps, aquatic traps, and road collecting.

The terrestrial drift fence and pitfall trap method has been described by Gibbons and Semlitsch (1982 [0804]) and used extensively at certain locations on the SRS. Sites where terrestrial drift fencing has been used for periods of two or more years are Pond C, Risher Pond, Ellenton Bay, the SREL Food Field, Karen's Pond, Lost Lake, Rainbow Bay, Sun Bay, Bullfrog Pond, and Flamingo Bay.

Aquatic trapping with large hoop nets and minnow traps has been used predominantly as a means of collecting aquatic species during intensive efforts to obtain information on turtle and salamander populations. Numerous other aquatic species besides turtles and salamanders have been incidentally trapped during these studies, providing a significant source of general information about their presence and relative abundance.

Road collecting has been a particularly rich source of specimens of both reptiles and amphibians and has yielded representatives of almost every species known from the SRS. Two primary reasons for the success of this technique are the numerous asphalt highways passing through relatively undisturbed habitats and the limited traffic flow during all times of the year and throughout most of the day and night.

Many other techniques have been used to collect reptiles and amphibians on the SRS for specific purposes or particular species; however, most have been described in literature resulting from SRS studies and do not warrant detailed description or discussion here. Local conditions and needs will dictate the most appropriate approach and techniques, and can be selected by the investigator at the time of study.

TAXONOMY

Herpetologists are not consistent in their choices of common names for some species of reptiles and amphibians. Therefore, we have tried to use those common names we think are most in vogue (see table 1). The simplest approach in most instances has been to adopt the names used by the Society for the Study of Amphibians and Reptiles (Collins et al., 1982) or given by Conant (1975). The choice of some species binomials is an even more difficult task. Again, we have done what we thought was best, realizing that some authorities may disagree. The important factor is, of course, to clarify which species is being considered, irrespective of what name we give it or how we spell the name.

The following list indicates some controversial taxonomic problems. The names we use in the book are followed by commonly used synonyms.

Table 1. Amphibians and Reptiles of the Savannah River Site

Order and suborder	Family	Scientific name	Common name
		CLASS AMPHIBIA	
Caudata (Salamanders)	Proteidae	*Necturus punctatus*	dwarf waterdog
	Amphiumidae	*Amphiuma means*	two-toed amphiuma
	Sirenidae	*Siren intermedia*	lesser siren
		Siren lacertina	greater siren
	Ambystomatidae	*Ambystoma maculatum*	spotted salamander
		Ambystoma opacum	marbled salamander
		Ambystoma talpoideum	mole salamander
		Ambystoma tigrinum	tiger salamander
	Salamandridae	*Notophthalmus viridescens*	eastern newt (red-spotted newt)
	Plethodontidae	*Desmognathus auriculatus*	southern dusky salamander
		*Desmognathus fuscus	dusky salamander
		Eurycea cirrigera	two-lined salamander
		Eurycea longicauda	long-tailed salamander (three-lined salamander)
		Eurycea quadridigitata	dwarf salamander
		Plethodon glutinosus	slimy salamander
		Pseudotriton montanus	mud salamander
		Pseudotriton ruber	red salamander
Anura (Frogs and toads)	Pelobatidae	*Scaphiopus holbrooki*	eastern spadefoot toad
	Bufonidae	*Bufo quercicus*	oak toad
		Bufo terrestris	southern toad
		*Bufo woodhousei	Woodhouse's toad
	Hylidae	*Acris crepitans*	northern cricket frog
		Acris gryllus	southern cricket frog
		Hyla avivoca	bird-voiced treefrog
		Hyla chrysoscelis	Cope's gray treefrog
		Hyla cinerea	green treefrog
		Hyla femoralis	pine woods treefrog
		Hyla gratiosa	barking treefrog
		Hyla squirella	squirrel treefrog
		*Pseudacris brimleyi	Brimley's chorus frog
		Pseudacris crucifer	spring peeper
		Pseudacris nigrita	southern chorus frog
		Pseudacris ocularis	little grass frog
		Pseudacris ornata	ornate chorus frog
		*Pseudacris triseriata	striped chorus frog
	Microhylidae	*Gastrophryne carolinensis*	eastern narrow-mouthed toad
	Ranidae	*Rana areolata*	crawfish frog (Carolina gopher frog)
		Rana catesbeiana	bullfrog
		Rana clamitans	green frog (bronze frog)

Table 1. Continued

Order and suborder	Family	Scientific name	Common name
		Rana grylio	pig frog
		Rana palustris	pickerel frog
		Rana sphenocephala	southern leopard frog
		Rana virgatipes	carpenter frog
		CLASS REPTILIA	
Crocodilia	Alligatoridae	*Alligator mississippiensis*	American alligator
Chelonia (Turtles)	Chelydridae	*Chelydra serpentina*	common snapping turtle
	Kinosternidae	*Kinosternon bauri*	striped mud turtle
		Kinosternon subrubrum	eastern mud turtle
		Sternotherus odoratus	stinkpot
	Emydidae	*Chrysemys picta*	painted turtle
		Clemmys guttata	spotted turtle
		Deirochelys reticularia	chicken turtle
		Pseudemys concinna	river cooter
		Pseudemys floridana	Florida cooter
		Terrapene carolina	eastern box turtle
		Trachemys scripta	slider turtle
	Trionychidae	*Trionyx spiniferus*	spiny softshell turtle
Squamata Lacertilia (Lizards)	Iguanidae	*Anolis carolinensis*	green anole (chameleon)
		Sceloporus undulatus	eastern fence lizard
	Teiidae	*Cnemidophorus sexlineatus*	six-lined racerunner
	Scincidae	*Eumeces fasciatus*	five-lined skink
		Eumeces inexpectatus	southeastern five-lined skink
		Eumeces laticeps	broadheaded skink
		Scincella lateralis	ground skink
	Anguidae	*Ophisaurus attenuatus*	slender glass lizard
		Ophisaurus ventralis	eastern glass lizard
Serpentes (Snakes)	Colubridae	*Carphophis amoenus*	worm snake
		Cemophora coccinea	scarlet snake
		Coluber constrictor	racer (black racer)
		Diadophis punctatus	ringneck snake
		Elaphe guttata	corn snake
		Elaphe obsoleta	rat snake
		Farancia abacura	mud snake
		Farancia erytrogramma	rainbow snake
		Heterodon platyrhinos	eastern hognose snake
		Heterodon simus	southern hognose snake
		Lampropeltis getulus	common kingsnake

Continued

Table 1. Continued

Order and suborder	Family	Scientific name	Common name
	Colubridae (cont'd)	*Lampropeltis triangulum*	scarlet kingsnake
		Masticophis flagellum	coachwhip
		Nerodia cyclopion	green water snake
		Nerodia erythrogaster	red-bellied water snake
		Nerodia fasciata	banded water snake
		*Nerodia sipedon	northern water snake
		Nerodia taxispilota	brown water snake
		Opheodrys aestivus	rough green snake
		Pituophis melanoleucus	pine snake
		Regina rigida	glossy crayfish snake
		Regina septemvittata	queen snake
		Rhadinaea flavilata	pine woods snake
		Seminatrix pygaea	black swamp snake
		Storeria dekayi	brown snake
		Storeria occipitomaculata	red-bellied snake
		Tantilla coronata	southeastern crowned snake
		Thamnophis sauritus	eastern ribbon snake
		Thamnophis sirtalis	common garter snake
		Virginia striatula	rough earth snake
		Virginia valeriae	smooth earth snake
	Elapidae	*Micrurus fulvius*	eastern coral snake
	Viperidae (Crotalidae)	*Agkistrodon contortrix*	copperhead
		Agkistrodon piscivorus	cottonmouth
		Crotalus horridus	timber rattlesnake (canebrake rattlesnake)
		Sistrurus miliarius	pygmy rattlesnake

Common names are those given by Conant (1975) or Collins et al. (1982), or are names applied locally. Those marked with an asterisk have been reported once from the SRS without subsequent verification or are suspected to be present but are difficult to distinguish from a closely related species. These are discussed under Problem Species (p. 94).

Our choice of a particular binomial does not indicate a taxonomic or systematic stand but does represent a name accepted by many authorities.

Usage in Book	Recent Synonyms
Eurycea cirrigera	*Eurycea bislineata*
Eurycea longicauda	*Eurycea guttolineata*
Pseudacris ocularis	*Limnaeodus ocularis*
	Hyla ocularis
Pseudacris crucifer	*Hyla crucifer*
Rana sphenocephala	*Rana utricularia*
Sternotherus odoratus	*Kinosternon odoratum*
Trachemys scripta	*Pseudemys scripta*
	Chrysemys scripta
Pseudemys concinna	*Chrysemys concinna*
Pseudemys floridana	*Chrysemys floridana*
Trionyx spiniferus	*Apalone spiniferas*
Heterodon platyrhinos	*Heterodon platirhinos*
Nerodia cyclopion	*Natrix cyclopion*
Nerodia erythrogaster	*Natrix erythrogaster*
Nerodia fasciata	*Natrix fasciata*
Nerodia sipedon	*Natrix sipedon*
Nerodia taxispilota	*Natrix taxispilota*
Regina rigida	*Natrix rigida*
Regina septemvittata	*Natrix septemvittata*

Key to the Orders of Amphibians and Reptiles

The following standard measurements are used: CL = carapace length; PL = plastron length; SVL = snout–vent length; TL = total length.

1 Scales absent; skin smooth in most species; if limbs present, no claws
 ....Class Amphibia, 2
1 Skin with scales or plates present; if limbs present, claws present....
 .... Class Reptilia, 4
  2 Tail absent .... Order Anura (frogs and toads), p. 39
  2 Tail present 3
3 Body elongate; forelimbs present; aquatic or terrestrial ....
 ....Order Caudata (salamanders), p. 28
3 Body globular; limbless or only hind limbs visible; strictly aquatic ...
 ....larval Anura (tadpoles), p. 43
  4 With bony or leathery shell ....Order Chelonia (turtles), p. 50
  4 Without bony or leathery shell 5
5 Cloacal opening a longitudinal slit; limbs present ....
 .... Order Crocodilia (alligators), no key
5 Cloacal opening a transverse slit; limbs present or absent ....
 ....Order Squamata (snakes and lizards), 6
  6 No eyelids or external ear opening; limbs absent ....
 ....Suborder Serpentes (snakes), p. 54
  6 Eyelids and external ear openings present; limbs present or absent
 ....Suborder Lacertilia (lizards), p. 52

Key to the Salamanders

1 No hind limbs; two forelimbs only; external gills present 2
1 Forelimbs and hind limbs present; external gills present or absent .. 3
   2 Costal grooves 31–34 (figure 2); total length seldom reaching 20 cm; solid color .... *Siren intermedia* (lesser siren)
   2 Costal grooves 36–39; total length often above 20 cm; yellow flecks on back and sides .... *Siren lacertina* (greater siren)
3 Nasolabial groove present, although may be difficult to see without magnification; external gills absent (figure 2) 4
3 Nasolabial groove absent or, if present, external gills also present . 11
   4 Four toes on hind foot; dark dorsolateral stripes; often a mid-dorsal row of small dark spots; 14–17 costal grooves; seldom reaching 5 cm in total length ....
    .... *Eurycea quadridigitata* (dwarf salamander)
   4 Five toes on hind foot 5
5 Body red or reddish brown 6
5 Body yellow, brown, or black 7

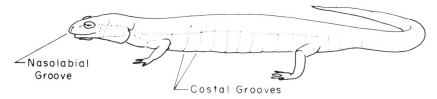

Figure 2. Position of costal grooves (between limbs) on typical salamander and nasolabial groove of plethodontid salamanders. The nasolabial groove may be difficult to see without magnification.

6 Body reddish brown or salmon color with distinct black spots; eye brownish .... *Pseudotriton montanus* (mud salamander)
6 Body reddish with muted dark specks; head with tiny white flecks; eye yellowish with brown stripe through center .... *Pseudotriton ruber* (red salamander)
7 White line behind eye 8
7 No white line behind eye 9
8 Usually with row of portholelike spots along side .... *Desmognathus auriculatus* (southern dusky salamander)
8 Usually without row of portholelike spots along side .... *Desmognathus fuscus* (dusky salamander)
9 Body black, flecked with silver, white, or brass; underside plain slate color with chin and throat dark; round tail .... *Plethodon glutinosus* (slimy salamander)
9 Body light with dark markings 10
10 Body yellow or tan with three dark longitudinal stripes; mid-dorsal stripe may be a series of dark spots; tail greater than one-half total length .... *Eurycea longicauda* (long-tailed salamander)
10 Two dark lines bordering light mid-dorsal stripe, often continuing to tip of tail; row of light circular spots along side; venter light yellow; tail usually less than one-half body length .... *Eurycea cirrigera* (two-lined salamander)
11 Body eel-like; tiny legs appear too small for body; two toes on each foot; gill openings behind head .... *Amphiuma means* (two-toed amphiuma)
11 Body not eel-like; legs proportional to body; four to five toes on each foot 12
12 External gills absent 13
12 External gills present 18
13 Skin rough 14
13 Skin smooth 15
14 Adult: skin rough; costal grooves indistinct; olive green to olive brown dorsum with small red spots; yellow belly with small black specks. .... *Notophthalmus viridescens* (eastern newt)
14 Red eft stage: skin rough; dorsum bright orange to dull red with black-bordered red spots .... *Notophthalmus viridescens* (eastern newt)

15 Dorsum dark with yellow or orange spots, blotches, or other markings 16
15 Dorsum dark with light or white markings 17
   16 Dorsum and belly dark; yellow or orange spots arranged in two irregular rows down back ....
      .... *Ambystoma maculatum* (spotted salamander)
   16 Dorsum dark with yellow or yellowish-brown irregular splotches or bands; belly with olive yellow and dark markings ....
      .... *Ambystoma tigrinum* (tiger salamander)
17 Body dark with distinct white or silvery crossbands; costal grooves 11–12 .... *Ambystoma opacum* (marbled salamander)
17 Body black, brown, or gray with pale bluish flecks; juvenile has striped belly; costal grooves 10–11 ....
      .... *Ambystoma talpoideum* (mole salamander)
   18 Four toes on hind foot; no nasolabial groove; dorsum dark with pale speckling .... *Necturus punctatus* (dwarf waterdog)
   18 Five toes on hind foot or four toes on hind foot and nasolabial groove present; this is a larval salamander of any of several species; go to larval salamander key below.

Key to the Larval Salamanders

Adapted from Altig and Ireland (1984), with modifications for local species.

The lack of a key to the identification of salamander larvae in the southeastern United States has been a problem to biologists for many years. Although the key presented here does not include all southeastern species, we believe it can be used to identify the larvae of species known to occur on the SRS or any that might possibly be found in the future. With the exception of *Pseudobranchus striatus*, all species of salamanders that occur in the South Carolina coastal plain and have aquatic larvae are included in the key and in the drawings. Additionally, the key and drawings should be applicable to coastal plain regions of North Carolina and Georgia.

As modeled after Altig and Ireland (1984), the key is meant for practical identification, especially for larval species that use the same breeding sites or are anatomically similar. The illustrations are used for visual reference to key characters and represent the norm for the species. Ontogenetic variation is noted if extreme or when information is available. *Plethodon*

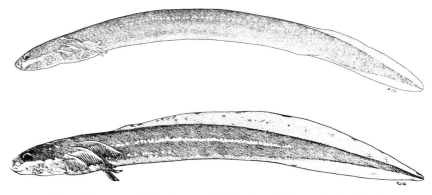

Plate 1. Larvae of *Siren intermedia* (top) and *S. lacertina* (bottom).

glutinosus is the only species of salamander known from the SRS that does not have an aquatic larval form.

1 Hind limbs or limb buds absent; four digits, usually with keratinized tips; body eel-like, 1–76 cm total length; larvae with black body and transparent fins; red or yellow head markings; body patterned with lichenlike markings; four fingers present; adults also aquatic, with gills 2
1 Hind limbs and toes present and completely differentiated; not marked as above 3
   2 Costal grooves 31–34 (axilla to vent); head markings on larvae red in life; dorsal fin usually not extending forward of mid-body; light mid-lateral stripe usually absent or not prominent .... *Siren intermedia* (lesser siren)
   2 Costal grooves 36–39; head markings in larvae yellow in life; dorsal fin extending forward of mid-body; usually with a prominent, light mid-lateral stripe .... *Siren lacertina* (greater siren)
3 Body elongate and cylindrical; one to three toes; legs appear disproportionately small for body; uniformly dark dorsally; one gill slit open; if gills present, thin, nonpigmented and transparent; two toes present .... *Amphiuma means* (two-toed amphiuma)
3 Body not elongate and cylindrical; four to five toes; legs appear normal in proportion to body; variously pigmented; one to four gill slits open; if gills present, as above or heavier and pigmented 4
   4 Four toes on hind foot 5

Plate 2. Larva of *Eurycea quadridigitata*.

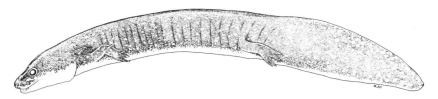

Plate 3. Larva of *Necturus punctatus*.

    4 Five toes on hind foot 7
5 Dorsal fin extends onto body; snout rounded; lungless; thin body; usually occurs in nonflowing water 6
5 Dorsal fin terminates on tail or at tail-body junction; lungs present; snout angular; stocky body; usually observed in flowing water 7
    6 Costal grooves 13–14; eye line present (occurs in South Carolina but not collected on the SRS) ....
................*Hemidactylium scutatum* (four-toed salamander)
    6 Costal grooves 14–17; striated pattern on lower side of body in some specimens; eye line absent ....
.... *Eurycea quadridigitata* (dwarf salamander)
7 Body long and slender; blunt or flattened snout; uniformly gray or brown dorsally; uniformly distributed light spots ....
.... *Necturus punctatus* (dwarf waterdog)
7 Not as above 8
    8 Grooves that delimit the labial folds on the lower jaw do not extend anteriorly to the mandibular symphysis; gills with rami and fimbriae throughout length; three to four gill slits open unless partially metamorphosed; tips of toes and soles of feet not keratinize 9

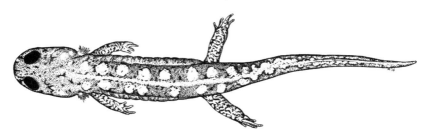

Plate 4. Larva of *Desmognathus auriculatus*.

8 Grooves that delimit the labial folds on lower jaw extend to and bisect the fold adjacent to the mandibular symphysis; gills without rami, short and branched from base, often appear silvery in life; four gill slits; gills bushy, heavily pigmented, and extending past front legs; toes and soles of feet often keratinized; dorsolateral series of circular spots prominent in smallest larvae (10–15 mm SVL) but faint to absent in large to nearly transformed individuals; dorsolateral black stripes absent; ventrolateral lateral line pores prominent and light; dorsal coloration drab and dark (*Desmognathus fuscus* somewhat similar in appearance but has not been reported from the SRS) ....
.... *Desmognathus auriculatus* (southern dusky salamander)
9 Dorsal fin extends onto body unless partially metamorphosed; lungs present; rarely observed in flowing water 10
9 Dorsal fin terminates on tail or at tail-body junction; lungs absent; most often observed in flowing water 16
   10 Four gill slits open unless partially metamorphosed, keratinized dental sheath absent; head somewhat pointed in dorsal view and not large; body slender; skin of larger specimens may be granular; greatest diameter of eye less than distance from eye to nostril; sides of head parallel; dark spots on tail ....
.... *Notophthalmus viridescens* (eastern newt)
   10 Three gill slits open unless partially metamorphosed, keratinized dental sheath usually present; broadly rounded head appears large on robust body; skin always smooth 11
11 Chin and/or throat heavily or lightly pigmented 12
11 Chin and/or throat unpigmented, cream colored 13

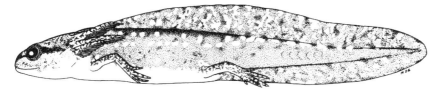

Plate 5. Larva of *Notophthalmus viridescens*.

12 Costal grooves 10–11; chin and ventral surface lightly and irregularly pigmented; longitudinal dark stripe on ventral surface in larvae 25 mm SVL; tail and body light with irregular dark bands except in large paedomorphs ....
.... *Ambystoma talpoideum* (mole salamander)

12 Costal grooves 11–13; throat evenly pigmented; ventral surface often pigmented; numerous light lateral spots; body and tail not banded .... *Ambystoma opacum* (marbled salamander)

13 Dorsum dark with conspicuous light, straight-edged stripes; yellow-gold mid-lateral stripe ragged edged but continuous from snout to mid-tail; head light brown with a dark stripe through eye; usually a clear area at base of dorsal fin (occurs in South Carolina but has not been collected on the SRS) ....
.... *Ambystoma cingulatum* (flatwoods salamander)

13 Not as above 14

14 Dorsum dark with light, ragged-edged stripes and numerous small, light flecks; light, mid-lateral stripe broken into series of spots; similar in appearance to *A. cingulatum* and *A. talpoideum* but has dark stripe through eye and no ventral stripe (occurs in South Carolina but has not been collected on the SRS) ....
.... *Ambystoma mabeei* (Mabee's salamander)

14 Not as above 15

15 Distinct dorsal irregular black spots; ground color highly mottled to uniformly gray; costal grooves usually more pigmented than costal folds; lateral light band and row of lateral spots on most larvae less than 60 mm total length; broad head relative to body length; toes flattened .... *Ambystoma tigrinum* (tiger salamander)

15 Not as above; head and toes rounded ....
.... *Ambystoma maculatum* (spotted salamander)

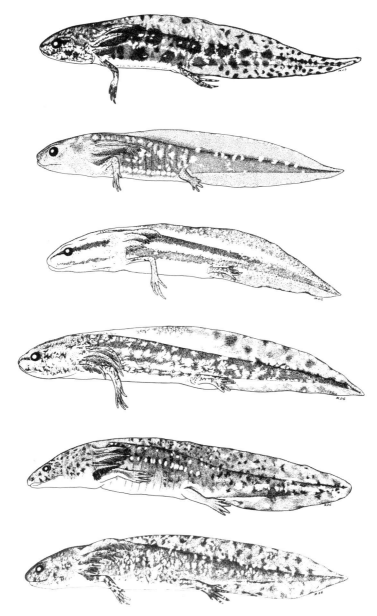

Plate 6. Larvae of (top to bottom) *Ambystoma talpoideum, A. opacum, A. cingulatum, A. mabeei, A. tigrinum, A. maculatum.*

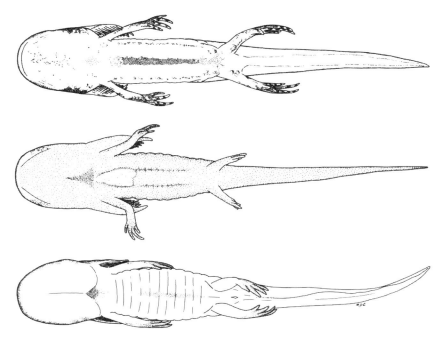

Plate 7. Ventral view of larvae of (top to bottom) *Ambystoma talpoideum*, *A. opacum*, *A. tigrinum*.

16 Costal grooves 18–19; six to nine costal folds between adpressed limbs; patterned much like adults (Conant,1975); head depressed; lateral line pores large and obvious (occurs in South Carolina but has not been collected on SRS) .... *Stereochilus marginatus* (many-lined salamander)
16 Not as above 17
17 Costal grooves 17–20; six or more costal folds between adpressed limbs; no indication of dorsal body stripe 18
17 Costal grooves 13–16; seven or fewer costal folds between adpressed limbs; dorsal body stripe usually visible 19
    18 Dorsum light brown with vermiculations; uniformly pigmented with fine-grained specks; usually without distinct spots; supraotic lateral line pores arranged in a circle; 16–17 costal grooves; eye stripe often present ....*Pseudotriton ruber* (red salamander)

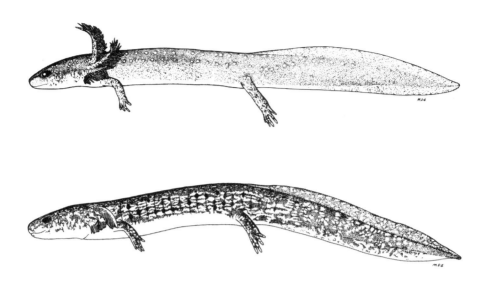

Plate 8. Larvae of *Pseudotriton ruber* (top) and *P. montanus* (bottom).

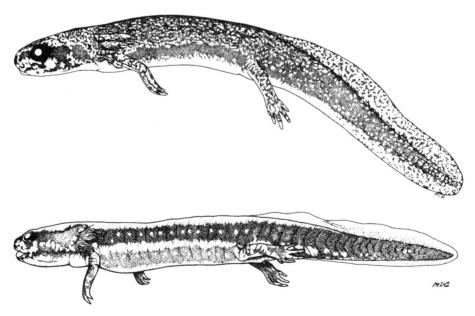

Plate 9. Larvae of *Eurycea cirrigera* (top) and *E. longicauda* (bottom).

18 Dorsum and sides uniformly light brown with small, distinct dark spots; flecks and reticulations along side of body .... *Pseudotriton montanus* (mud salamander)
19 Dorsum with longitudinal series of light spots 20
19 Dorsum without longitudinal series of light spots; dorsolateral dark stripes absent, or lateral body surface not dark .... *Eurycea cirrigera* (part) (two-lined salamander)
    20 Dark vertical bars on side of tail or this area uniformly dark; usually 13 costal grooves ....
    .... *Eurycea longicauda* (long-tailed salamander)
    20 Dark vertical bars absent on side of tail or this area not uniformly dark; usually 14 costal grooves ....
    .... *Eurycea cirrigera* (part) (two-lined salamander)

Key to the Frogs and Toads

1 Pupil of eye vertical; underside of hind foot with single dark, hard tubercle on heel ...... *Scaphiopus holbrooki* (eastern spadefoot toad)
1 Pupil of eye round; underside of hind foot with two dark, hard tubercles or with no tubercles 2
    2 Two tubercles on underside of foot; body dry and rough; paratoid glands present on top of head 3
    2 No tubercles; body smooth, usually moist; no paratoid glands 5
3 Adults never more than 40 mm SVL; background color gray with clearly defined white median stripe down back ....
.... *Bufo quercicus* (oak toad)
3 Adults reddish, light brown, or dark brown 4
    4 Parietal cranial crest ends in a conspicuous knob; dorsal spots usually have one to three warts .... *Bufo terrestris* (southern toad)
    4 Parietal cranial crest present but without a pronounced knob; dorsal spots often with more than three warts ....
.... *Bufo woodhousei* (Woodhouse's toad)
5 Tympanum absent; head pointed, with transverse fold behind eyes; no toe webbing; color gray or brownish; SVL less than 30 mm ......
.... *Gastrophryne carolinensis* (eastern narrow-mouthed toad)
5 Tympanum present; no transverse head fold; at least partial webbing between some toes 6
    6 Hind foot with minimal webbing; tips of toes expanded into small discs in most species; size usually less than 60 mm SVL 7
    6 Hind foot with extensive webbing; adult size usually greater than 50 mm SVL 20
7 Adults less than 17 mm SVL; dark stripe through eye ....
.... *Pseudacris ocularis* (little grass frog)

Key to the Frogs and Toads

7 Adults greater than 20 mm SVL 8
   8 Dark stripe on rear of thigh; dark triangle usually present between eyes 9
   8 No dark stripe on rear of thigh 10
9 Dark stripe on rear of thigh with ragged edges ....
.... *Acris crepitans* (northern cricket frog)
9 Dark stripe with no ragged edges ....
.... *Acris gryllus* (southern cricket frog)
   10 Tips of toes expanded only slightly, less than one-half diameter of tympanum 11
   10 Tips of toes larger than one-half diameter of tympanum 14
11 Very distinct unbroken black stripe from snout through eye to base of thigh; underside yellow, usually with dark spots ....
.... *Pseudacris brimleyi* (Brimley's chorus frog)
11 No distinct black stripe; underside not yellow 12
   12 Light-bordered dark spots or blotches on sides; body green, brown, or gray; adult SVL approximately 35 mm ....
.... *Pseudacris ornata* (ornate chorus frog)
   12 No blotches on sides 13
13 Dark triangle between eyes ....
.... *Pseudacris triseriata* (striped chorus frog)
13 No dark triangle between eyes ....
.... *Pseudacris nigrita* (southern chorus frog)
   14 Usually with a dark "X" across back; size less than 35 mm SVL
.... *Pseudacris crucifer* (spring peeper)
   14 No "X" on back 15
15 Rear of thigh with light-colored green or yellow wash, or orange spots 16
15 Rear of thigh without conspicuous colored markings 18
   16 Rear of thigh greenish; light-colored spot below each eye ....
.... *Hyla avivoca* (bird-voiced treefrog)
   16 Rear of thigh with yellow or orange markings; light-colored spot below eye present or absent 17
17 Light-colored spot below each eye; back rough or warty; rear of thigh yellowish .... *Hyla chrysoscelis* (Cope's gray treefrog)
17 No spot below eye; back smooth; distinct yellow-gold spots on rear of thigh .... *Hyla femoralis* (pine woods treefrog)
   18 Body bright solid green with no large spots, although bright yel-

low flecks may be present; usually with distinct yellowish stripe along sides .... *Hyla cinerea* (green treefrog)
18 No distinct clearly defined stripe along sides 19
19 Markings along side absent or inconspicuous; adult size less than 45 mm SVL .... *Hyla squirella* (squirrel treefrog)
19 Brown spots often present on green body; side markings may form diffuse stripe; size of adults greater than 50 mm .... *Hyla gratiosa* (barking treefrog)
    20 Dorsolateral ridges present; distinct spots may be present on back 21
    20 Dorsolateral ridges absent; spots inconspicuous or absent 24
21 No distinct spots; dorsolateral ridge incomplete, extending about two-thirds down back .... *Rana clamitans* (bronze frog)
21 Spots present; dorsolateral ridge complete 22
    22 Square spots in two parallel rows down back; orange or yellow markings in groin area .... *Rana palustris* (pickerel frog)
    22 Spots not square; no yellowish coloration on legs 23
23 Snout pointed; belly usually light in color .... *Rana sphenocephala* (southern leopard frog)
23 Snout bluntly rounded; body chunky in appearance; belly usually heavily pigmented .... *Rana areolata* (crawfish frog)
    24 Four distinct longitudinal lines down back .... *Rana virgatipes* (carpenter frog)
    24 Longitudinal lines absent or (on small *Rana grylio*) pale and indistinct 25
25 Belly dark with light markings; light spots on lips (not reported from SRS) .... *Rana heckscheri* (river frog)
25 Belly light with dark markings 26
    26 Web on hind feet extending to tip of the fourth (longest) toe; snout pointed .... *Rana grylio* (pig frog)
    26 Web extending only partially along fourth (longest) toe; snout blunt .... *Rana catesbeiana* (bullfrog)

Key to the Tadpoles

Adapted from Altig (1970), with permission of the author.

1 Jaws without keratinized sheaths; oral disc and labial teeth absent (figure 3); dorsum dark brown to black; tail stripe usually distinct; underside of labial flap without excrescences; ratio of total length to body length 2.1 or more (figure 4); body round from dorsal view
.... *Gastrophryne carolinensis* (eastern narrow-mouthed toad)
1 Jaws with keratinized sheaths; oral disc and labial teeth present ... 2
    2 Anal tube medial; eyes dorsal (figure 5) 3
    2 Anal tube dextral; eyes dorsal or lateral 4
3 Papillary border with a wide dorsal gap about equal to A-1 and a ventral gap equal to or larger than P-3; oral disc emarginate; labial tooth-row formula 1-2(2)/3(1) (figure 3) 5
3 Papillary border without a ventral gap, dorsal gap present or not; oral disc not emarginate; body somewhat depressed, typically wider posteriorly than anteriorly; to 35 mm total length; dorsum usually dark brown to black; jaws narrow; labial tooth-row formula 4-6(2-6)/3-6(1-3); spiracle equidistant between eye and vent; ratio of interorbital distance to internarial distance 1.8 or less; ratio of tail height to musculature height 2.5 or less; last posterior tooth row longer than upper jaw and one-half or more times next anterior row; A-2 normally with a median gap ....
.... *Scaphiopus holbrooki* (eastern spadefoot toad)
    4 Eyes lateral or dorsal; oral disc not emarginate; papillary border commonly reaches considerable distance above lateral tips of A-1; labial tooth-row formula 2(2)/2-4(1) 7
    4 Eyes dorsal; oral disc emarginate; papillary border does not reach or barely reaches above lateral tips of A-1; labial tooth-row formula 1-7(2-7)/2-6(1), commonly 2-3/3-4; labial tooth-row formula rarely 2/2; papillary border without a posterior gap; marginal papillae common 27
5 Papillary border extends to lateral tips of P-2, although a single papilla may occur at each end of P-3; P-1 with a median gap; P-3 short; tail musculature bicolored and often with dorsal light saddles ......
.... *Bufo quercicus* (oak toad)
5 Not as above 6

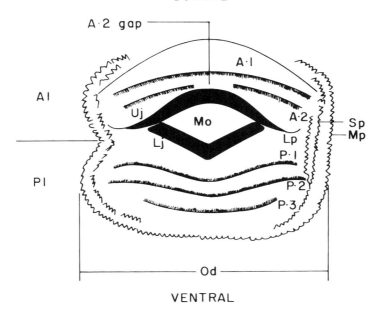

Figure 3. Frontal view of tadpole mouthparts. The animal's right side is emarginate; its left side is not emarginate. Al, anterior labium; A-1, first anterior tooth row; A-2, second anterior tooth row; Lj, lower jaw; Lp, lateral process of upper jaw; Mo, mouth; Mp, marginal papilla; Od, oral disc; Pl, posterior labium; P-1, first posterior tooth row; P-2, second posterior tooth row; P-3, third posterior tooth row; Sp, submarginal papilla; Uj, upper jaw. The labial tooth-row formula is standardized as follows: the numeration indicates the number of rows on the anterior labium; the denominator indicates the number on the posterior labium; numbers in the parentheses indicate row numbers having gaps in the middle.

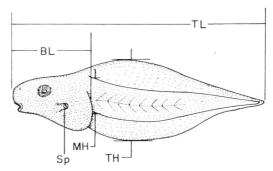

Figure 4. Body size measurements of tadpole. BL = body length; TL = total length; SP = spiracle opening; TH = tail height; MH = musculature height.

6 Upper and lower fins equal to musculature height; dorsal fin often higher than ventral; dorsum dark brown to black, with a light oblique mark behind each eye in life ....
.... *Bufo terrestris* (southern toad)
6 Upper and lower fins lower than musculature height; fins subequal in height; dorsum dark, commonly with light mottling in life; snout rounded in lateral view; eyes large; ratio of tail length to tail height 2.8 or more; musculature often not distinctly bicolored; ratio of tail height to musculature height 2 or more ....
.... *Bufo woodhousei* (Woodhouse's toad)
7 Two rows of labial teeth on posterior labium 8
7 Three rows of labial teeth on posterior labium 12

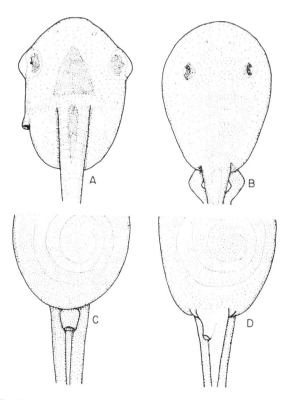

Figure 5. Tadpole morphological patterns indicating lateral (A) and dorsal (B) eyes and medial (C) and dextral (D) vent.

8 A-2 gap wide; spiracular tube at least partially free from body wall; tail tip often black; dorsum of tail musculature frequently banded; body slightly depressed; eyes dorsolateral; nostrils large; fins without bold markings 9

8 A-2 gap narrow; spiracular tube almost fully attached to body wall; tail tip never black; dorsum of tail musculature banded or not; body globular; eyes lateral; nostrils large or small; fins with or without bold markings 10

9 Free section of spiracular tube long, almost entire length of tube; throat dark; tail musculature finely flecked ....
.... *Acris gryllus* (southern cricket frog)

9 Free section of spiracular tube short, one-half or less the length of the tube; throat light, tail musculature mottled or reticulated ....
.... *Acris crepitans* (northern cricket frog)

10 Tail musculature striped in lateral view; light stripe extends forward to eye from dorsal musculature stripe; throat and chest mottled; dorsum of tail musculature often banded; small size ....
.... *Pseudacris ocularis* (part) (little grass frog)

10 Tail musculature not striped in lateral view; light stripes from tail to eye absent; throat and chest light; dorsum of tail not banded; larger size 11

11 Tail musculature unicolored or bicolored; one row of marginal papillae; fins clear with some stellate melanophores; A-2 equal to or less than A-1 .... *Pseudacris triseriata* (part) (striped chorus frog)

11 Tail musculature mottled; two rows of marginal papillae; fins clear, with or without large black blotches; if blotches present, a clear area near musculature; A-2 longer than A-1 ....
.... *Pseudacris crucifer* (part) (spring peeper)

12 Papillary border with a posterior gap 13
12 Papillary border without a posterior gap 14

13 Tail musculature striped in lateral view; light stripes extend to eye from dorsal light stripe on tail; dorsum of tail musculature often banded; to 25 mm total length; fins clear or with few melanophores
.... *Pseudacris ocularis* (part) (little grass frog)

13 Tail musculature not striped, mottled in lateral view; light stripes from tail to eyes absent; dorsum of tail musculature not banded; to 35 mm total length; fins often with large dark blotches and a clear area near musculature .... *Pseudacris crucifer* (part) (spring peeper)

14 Tail musculature dark with light (reddish in life) dorsal saddles; fins dark without bold markings; light interorbital and orbitonasal bands present in life; gut not visible, belly black .... *Hyla avivoca* (bird-voiced treefrog)

14 Tail musculature dark or light, without dorsal saddle; tail fins variously patterned; interorbital bars absent, orbitonasal bars may or may not be present; gut visible or not 15

15 P-3 long, 0.65 or more times P-2 and longer than upper jaw; submarginal papillae well developed 16

15 P-3 short, less than 0.65 times P-2 and equal to or shorter than upper jaw; submarginal papillae poorly developed to absent 18

    16 Tail musculature distinctly striped; fins flecked or blotched with clear area remaining near musculature; clear part of fin reddish in large specimens in life; flagellum well developed and clear of pigment .... *Hyla femoralis* (pine woods treefrog)

    16 Tail musculature not distinctly striped; fins blotched or not, with or without clear area near musculature; clear part of fin reddish or not; flagellum less developed 17

17 Ratio of tail height to body height at eye level 1.5 or less; dorsal fin equal to or greater than musculature height; ratio of tail length to tail height 3.1 or more; ratio of greatest body width to basal width of tail musculature 2.7 or more; large specimens often with reddish fins; throat seldom pigmented .... *Hyla chrysoscelis* (Cope's gray treefrog)

17 Ratio of tail height to body height at eye level 1.5 or more; dorsal fin less than musculature height; ratio of tail length to tail height 3.3 or less; ratio of greatest body width to basal width of tail musculature 2.7 or less; large specimens without red fins; throat typically pigmented .... *Hyla squirella* (squirrel treefrog)

    18 Tail musculature striped in lateral view 19

    18 Tail musculature bicolored, unicolored, or mottled in lateral view 20

19 Wide lateral tail stripe bordered dorsally and ventrally by light stripes; dorsal light stripe extends forward to eye; dorsum of tail musculature often banded; to 25 mm total length .... *Pseudacris ocularis* (part) (little grass frog)

19 Dark stripe with indistinct edges projects obliquely from center of tail musculature to dorsal margin of musculature at about mid-length; stripe to eye absent; dorsum of tail not banded; to 35 mm total length (not reported from SRS) .... *Hyla andersoni* (pine barrens treefrog)

20 Jaws wide; upper jaw angulate; dorsal fin high, terminating anterior to spiracle; body compressed slightly; to 55 mm total length; tail clear, black, or mottled 21

20 Jaws narrow to medium; upper jaw not angulate; dorsal fin high or not; tail clear or blotched; always less than 55 mm total length 22

21 Total length less than 30 mm; fins clear; light stripe extends from tail musculature forward to eye; dorsum of tail musculature with a black saddle slightly anterior to mid-body .... *Hyla gratiosa* (part) (barking treefrog)

21 Total length more than 30 mm; fins clear, black, or mottled; stripe from tail to eye absent; black saddle on tail musculature absent .... *Hyla gratiosa* (part) (barking treefrog)

22 Fins and tail musculature typically mottled or reticulated without a clear area near musculature; A-2 gap ratio 3 or more; light orbitonasal stripe present at least in life; small specimens with two light body blotches that form an incomplete transverse body band (usually lost in preservative); dorsal fin height equals ventral .... *Hyla cinerea* (green treefrog)

22 Fins not mottled or reticulated, sometimes blotched; if blotched, a clear area present near musculature; A-2 gap ratio 3 or less; light orbitonasal stripe absent; light body blotches absent; dorsal fin variable 23

23 Fins commonly blotched with a clear area remaining near musculature; tail musculature mottled; dorsal fin higher than ventral; P-3 very short .... *Pseudacris crucifer* (part) (spring peeper)

23 Fins not blotched, either clear or with a few stellate melanophores; tail musculature mottled, unicolored, bicolored, or striped; dorsal fin variable 24

24 Throat pigmented; A-2 gap relatively wide; tail musculature striped, and dorsal light stripe extends forward to eye; P-1 with a median gap; to 21 mm .... *Pseudacris brimleyi* (Brimley's chorus frog)

24 Throat not pigmented; A-2 relatively narrow; tail musculature not striped; no light stripe from tail to eye; P-1 with or without median gap; to 35 mm total length 25

25 Dorsal fin high, extending anterior of spiracle; tail musculature distinctly bicolored; P-1 indented or with a narrow median gap; ratio of tail height to musculature height 3.5 or more ....
............................*Pseudacris ornata* (ornate chorus frog)
25 Dorsal fin high or low, usually not extending past spiracle; tail musculature unicolored or indistinctly bicolored, i.e., not for full length of tail or with considerable pigment in lower half; P-1 with or without a median gap; ratio of tail height to musculature height 3.2 or less
.... 26
  26 Chest pigmented; dorsum uniform black to dark brown without small black dots; dorsal fin height less than tail musculature .....
.......................*Pseudacris nigrita* (southern chorus frog)
  26 Chest not pigmented; dorsum black to dark brown and typically with small black dots; dorsal fin height about equal to tail musculature .... *Pseudacris triseriata* (striped chorus frog)
27 Tail and body greenish, unicolored or more commonly patterned with distinct black dots; fins similarly patterned with more dots in dorsal fin than in ventral; venter clear to white depending on size, with or without a contrasting pattern; proximal portion of tail often more opaque than remainder; small live specimens (less than 25 mm total length) black with transverse gold bands on snout and body; tail appears bicolored due to pigment around caudal blood vessels .....
.... *Rana catesbeiana* (bullfrog)
27 Variously patterned; venter clear, white or dark, seldom with a contrasting pattern 28
  28 Lower jaw wide; nostrils medium sized; skin thin; gut usually visible 29
  28 Lower jaw narrow; nostrils small to medium; skin thick or not; gut usually not visible 31
29 A-2 gap ratio 2 or more; marginal papillae below P-3 large, 10 or fewer present; fins usually heavily marked, often with dark suffusion; ratio of P-1 to P-3 1.3 or more; gut often only slightly visible ....
..................................*Rana palustris* (pickerel frog)
29 A-2 ratio less than 2; marginal papillae below P-3 small, more than 10 present; fins heavily marked or not, speckled or spotted; ratio of P-1 to P-3 1.5 or less, gut visible or not 30
  30 Gut usually visible; unpigmented throat patch with contrasting margins often present; if tail marked, usually not with large spots;

dorsum not stippled; dorsal fin rounded; keratinized areas at medial tips of P-1 absent ....
.... *Rana sphenocephala* (southern leopard frog)
30 Gut visible or not; throat unpigmented without contrasting margins or evenly pigmented; if tail marked, usually with large spots; dorsum often appears stippled; dorsal fin frequently triangular; keratinized areas at medial tips of P-1 present on large specimens
.... *Rana areolata* (crawfish frog)
31 Stripe in dorsal fin and stripe on tail musculature absent; light spots surrounded by dark pigment at fin edges absent or indistinct; gut visible or not; A-2 gap ratio 5 or more; dorsum greenish; row of submarginal papillae absent between P-3 and marginal papillae ....
.... *Rana clamitans* (bronze frog)
31 Stripe or row of dots formed by pigment around lateral line pores present in dorsal fin and a less prominent stripe usually present on tail musculature; light spots surrounded by dark pigment present near edge of fin; gut slightly or not visible 32
32 Dorsum brown with black dots; venter brown in preservative, yellow to buff in life; tail musculature stripe typically present; gut slightly or not visible; row of submarginal papillae present between P-3 and marginal papillae; ratio of tail length to tail height 2 or less .... *Rana virgatipes* (carpenter frog)
32 Dorsum brownish to greenish with black dots or mottled; venter white, often with a contrasting pattern; tail musculature stripe typically indistinct to absent; gut slightly or not visible; row of submarginal papillae absent between P-3 and marginal papillae; ratio of tail length to tail height 2.2 or more ....
.... *Rana grylio* (pig frog)

Key to the Turtles

1 Shell leathery, flat, and light brown in color; similar to a pancake in appearance .... *Trionyx spiniferus* (spiny softshell turtle)
1 Shell bony and dome-shaped 2
   2 Four toes on hind foot; plastron large with one transverse hinge; capable of sealing entire shell; yellowish or orange spots usually present on head, limbs, and shell ....
     .... *Terrapene carolina* (eastern box turtle)
   2 Five toes on hind foot 3
3 Plastron with two transverse hinges; pectoral scutes triangular 4
3 Plastron with one or no transverse hinge; pectoral scutes quadrangular 5
   4 Head and carapace black; if yellow head stripes visible, they do not extend from eye to snout ....
     .... *Kinosternon subrubrum* (eastern mud turtle)
   4 Faint yellow longitudinal lines sometimes visible on head or carapace; head stripes nearly always extend from eye ....
     .... *Kinosternon bauri* (striped mud turtle)
5 Tail greater than one-half the carapace length and sawtoothed along upper side; barbels on chin; head without yellow stripes ....
   .... *Chelydra serpentina* (common snapping turtle)
5 Tail less than one-half the carapace length; with or without barbels on chin; head with yellow stripes 6
   6 Plastron with one inconspicuous transverse hinge; barbels on chin and throat; shell solid black or brown with no yellow markings ....
     .... *Sternotherus odoratus* (stinkpot)
   6 Plastron without a hinge; no barbels; shell usually with yellow markings 7

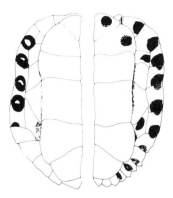

Figure 6. Plastrons of *Pseudemys concinna* or *P. floridana* (left) and *Trachemys scripta* (right) showing distinction between marginal spotting. On *P. concinna*, the midline of the plastron is also dark.

7 Carapace, head, and neck black with scattered, conspicuous yellow or orange dots .... *Clemmys guttata* (spotted turtle)
7 Carapace without conspicuous yellow or orange dots; head and neck with yellow stripes 8
   8 Carapace dark olive to black; without yellow markings on underside of marginals ....*Chrysemys picta* (painted turtle)
   8 Carapace greenish brown, dark olive, or black; yellow markings on underside of marginals and usually on carapace 9
9 Carapace with concentric light and dark markings on scutes and figure "C" on the second costal scute; dark spots with light centers along underside of all or most marginals (figure 6); dark plastral

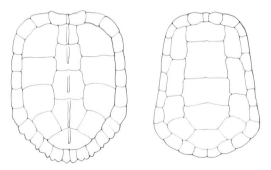

Figure 7. Carapace shapes of *Trachemys scripta* (left) and *Deirochelys reticularia* (right).

       markings present, at least down midline ....
       .... *Pseudemys concinna* (river cooter)
   9 Carapace without a "C" on the second costal scute 10
      10 Skin between hind legs marked irregularly with yellow and black;
         dark spots with light centers on underside of anterior marginals;
         posterior marginals without spots; dark plastral markings absent
         .... *Pseudemys floridana* (Florida cooter)
      10 Skin between hind legs marked with uniform vertical yellow and
         black stripes 11
  11 Broad yellow stripe on forelimbs; carapace distinctly longer than
     wide (figure 7); head with yellow stripes ....
     .... *Deirochelys reticularia* (chicken turtle)
  11 Narrow yellow stripes on forelimbs; yellow blotch behind eye; carapace round or nearly so .... *Trachemys scripta* (slider turtle)

Key to the Lizards

1 Limbs absent 2
1 Limbs present 3
  2 No dark horizontal stripes below lateral groove; no distinct dark mid-dorsal stripe .... *Ophisaurus ventralis* (eastern glass lizard)
  2 Narrow dark horizontal stripes below lateral groove; dark mid-dorsal stripe or broken stripe ....
.... *Ophisaurus attenuatus* (slender glass lizard)
3 Dorsal scales shiny and smooth 4
3 Dorsal scales not shiny; scales granular or keeled 7
  4 Transparent disc in lower eyelid; dorsal color brown without conspicuous light stripes; belly white or yellowish; size small, SVL never exceeding 4 cm .... *Scincella lateralis* (ground skink)
  4 No transparent disc in lower eyelid; if less than 4 cm SVL, then conspicuous light stripes present on black body; tail of all small and some large individuals blue 5
5 Scales on underside of tail uniform in size (figure 8); five light dorsal stripes, the mid-dorsal one narrow ....
.... *Eumeces inexpectatus* (southeastern five-lined skink)

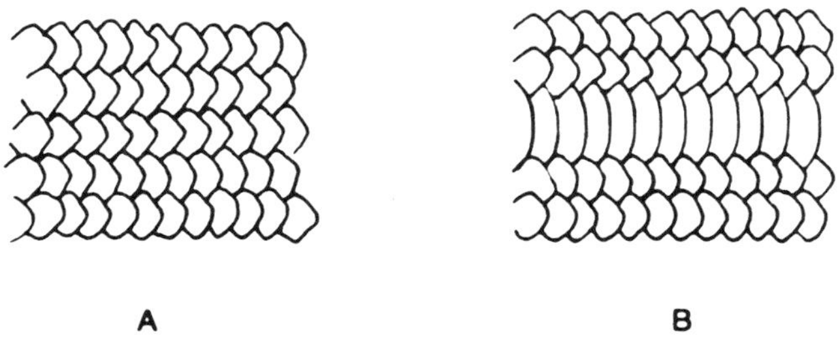

        A   B

Figure 8. Underside of tail indicating scales uniform (A) as in *Eumeces inexpectatus*, and medial scales wider than long (B) as in *E. fasciatus* and *E. laticeps*.

5 Scales on underside of tail not uniform in size 6
   6 Five upper labials in front of subocular scale; no enlarged postlabials (figure 9) ...... *Eumeces laticeps* (broadheaded skink)
   6 Usually four upper labials in front of subocular scale; two enlarged postlabials .... *Eumeces fasciatus* (five-lined skink)
7 Ventral scales rectangular in eight longitudinal rows; six light longitudinal stripes dorsally .... *Cnemidophorus sexlineatus* (six-lined racerunner)
7 Ventral scales not rectangular and not in eight longitudinal rows; no stripes 8
   8 Dorsal scales granular and pointed; color grayish .... *Sceloporus undulatus* (eastern fence lizard)
   8 Dorsal scales granular and not pointed; toe pads present; dorsal color variable green to brown or grayish, sometimes with lighter dorsal stripe .... *Anolis carolinensis* (green anole)

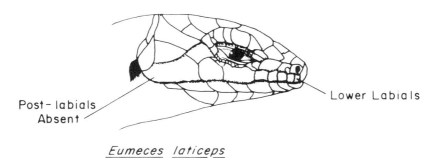

Eumeces laticeps

Figure 9. Labial scale pattern of *E. laticeps* showing right upper labials (five anterior to the subocular) and no postlabial scales. *E. fasciatus* has seven upper labials and two postlabials.

Key to the Snakes

Common names given in boldface type indicate venomous snakes.

1 Pit between eye and nostril; many or most scales on underside of tail in a single row (figure 10) 2
1 No pit between eye and nostril; scales on underside of tail in double row 5
   2 Top of head with many small scales (figure 11); rattle or button on tail; most of tail velvety black (figure 12); basic color gray or tan with black, chevronlike bands across back; pinkish to yellow stripe down center of back .... *Crotalus horridus* (**canebrake rattlesnake**)
   2 Top of head with nine large scales; rattles or button present or absent 3
3 Rattles or button on tail; size small, seldom exceeding 60 cm; basic color gray with dark blotches along back and sides ....
.... *Sistrurus miliarius* (**pygmy rattlesnake**)
3 No rattles or button; basic color black, brown, or reddish 4
   4 Dorsal scale rows 23 at mid-body (figure 13); basic color light brown or reddish with darker crossbands, often hourglass shaped, along body .... *Agkistrodon contortrix* (**copperhead**)
   4 Dorsal scale rows 25 at mid-body; basic color brown to dark brown or black in specimens over 45 cm long; sometimes reddish in smaller specimens ...... *Agkistrodon piscivorus* (**cottonmouth**)
5 Scale rows at mid-body 13-19 6
5 Scale rows at mid-body 21 or more 26
   6 All scales smooth (figure 14) 7
   6 All scales distinctly keeled 19
7 Basic color uniformly black or light brown above, except for head and neck region in some instances 8

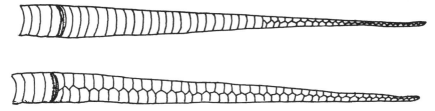

Figure 10. Underside of tail of poisonous crotalid (top; single row posterior to vent) and nonvenomous colubrid (bottom; double row posterior to vent).

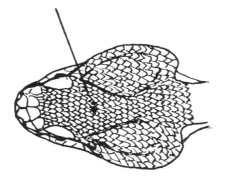

Figure 11. Head of *Crotalus*, showing numerous small scales.

Figure 12. Solid black tail of *Crotalus horridus*.

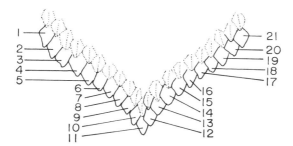

Figure 13. Counting system for dorsal scale rows.

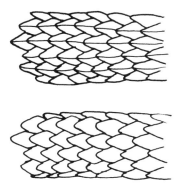

Figure 14. Comparison of keeled (top) and smooth (bottom) scales.

7 Not uniformly black or brown above, with some form of markings on back (blotches, rings, stripes) or with basic color grading from black head to light brown tail 14
   8 Basic color tan to light brown; length seldom exceeding 30 cm ... 9
   8 Basic color black or dark gray 11
9 Head black, with a white band and black ring around neck ....
.................*Tantilla coronata* (southeastern crowned snake)
9 Body uniform brown above from head to tail 10
   10 Scale rows at mid-body 15; belly whitish or grayish ....
.... *Virginia valeriae* (smooth earth snake)
   10 Scale rows at mid-body 13; belly pinkish ....
..........................*Carphophis amoenus* (worm snake)
11 Shiny black above; dark gray to black belly; white chin; no red or yellow color on any part of body ...... *Coluber constrictor* (black racer)
11 Black above; belly yellow, orange, or red 12
   12 Dull, velvety black above; yellow ring around neck; yellow or orange belly usually with a single row of black spots medially; length seldom exceeding 30 cm ....
.... *Diadophis punctatus* (ringneck snake)
   12 Uniform black above from snout to tail; belly with red or orange
.... 13
13 Belly solid orange or red; length seldom exceeding 50 cm ....
..........................*Seminatrix pygaea* (black swamp snake)
13 Belly checkerboard in appearance, alternating bright red and black
.... *Farancia abacura* (mud snake)

14 Body color including red, orange, or yellow 15
14 Body color not including red, orange, or yellow 17
15 Body shiny black above with three thin red stripes; belly red, orange, or yellow with black spots .. *Farancia erytrogramma* (rainbow snake)
15 Body with transverse bands or rings of red, black, and yellow or white 16
    16 Body rings continuing across belly; encircling body; snout black .... *Micrurus fulvius* (**eastern coral snake**)
    16 Belly white or gray; snout red ....
      .... *Cemophora coccinea* (scarlet snake)
17 Head and front half of body dark brown to velvety black; posterior half of body brown .... *Masticophis flagellum* (coachwhip)
17 Head not black; length usually not exceeding 50 cm 18
    18 Row of dark gray or brownish blotches down back; small dark spots on sides and belly ....
      .... *Coluber constrictor* (juvenile black racer)
    18 Light brown or tan above with thin dark crossbands; eyes noticeably larger than those of other snakes ....
      .... *Masticophis flagellum* (juvenile coachwhip)
19 Body slender, solid green above; belly yellow ....
  .... *Opheodrys aestivus* (rough green snake)
19 Body not solid green above 20
    20 Usually with three light yellow or greenish stripes down center of back and along side; belly yellowish, light greenish, or gray 21
    20 Body black or brownish above without three yellow stripes 22
21 Side stripes confined to scale rows 3 and 4 on each side (figure 15); lip scales yellowish without dark markings; lip scales not separated by dark markings .... *Thamnophis sauritus* (eastern ribbon snake)
21 Side stripes, when present, confined to scale rows 2 and 3 on each side; lip scales separated by dark markings ....
    .... *Thamnophis sirtalis* (common garter snake)
    22 Body dark or light brown above, belly red or light colored without stripes; scale rows 15 or 17 23
    22 Body black or dark brown above; belly white or yellowish with two or four dark stripes or rows of dots; scale rows 19 25
23 Belly red or orange; scale rows 15 ....
  .... *Storeria occipitomaculata* (red-bellied snake)
23 Belly grayish or brownish; scale rows 17 24

58 *Guide to the Reptiles and Amphibians of the SRS*

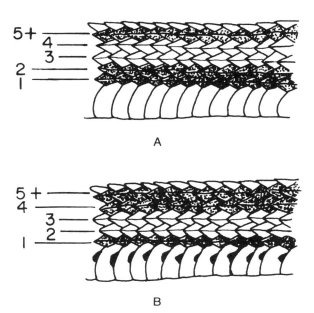

Figure 15. Lateral line scale patterns of *Thamnophis sauritus* (top) (yellow line on third and fourth scale rows) and *T. sirtalis* (bottom).

24 Body color mostly solid brown to gray with few or no markings; nose pointed; sometimes with pale band at neck .... *Virginia striatula* (rough earth snake)
24 Body color light brown or gray with darker crossbands, spots, or checkerboard markings on back and sides; nose rounded .... *Storeria dekayi* (brown snake)
25 Two distinct stripes or rows of spots down belly; lower sides lacking light stripes .... *Regina rigida* (glossy crayfish snake)
25 Four stripes or rows of spots down belly; light stripe along lower sides .... *Regina septemvittata* (queen snake)
26 All scales smooth; color pattern includes yellow rings encircling body 27
26 Scales keeled (weakly keeled only on sides in some forms); color dark or patterned with bands or blotches but not with yellow crossbands or rings 28

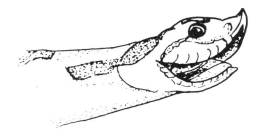

Figure 16. Upturned rostral scale of *Heterodon simus*.

27 Body completely encircled by red, black, and yellow rings; red snout; length seldom exceeding 60 cm .... *Lampropeltis triangulum* (scarlet kingsnake)
27 Body shiny black with yellow or white rings or crossbands; no red present; length often exceeds 60 cm .... *Lampropeltis getulus* (common kingsnake)
   28 Basic color light gray to reddish tan above with dark blotches; entire belly and underside of tail white with no markings .... *Pituophis melanoleucus* (pine snake)
   28 Basic color patterns darker than light gray; parts of underside dark or patterned 29
29 Rostral scale (at tip of nose) distinctly pointed or upturned (figure 16) 30
29 Rostral scale rounded 31
   30 Underside of tail darker than or same shade as belly; patterned above with dark blotches on light tan or brownish body; body never black; length seldom exceeding 50 cm .... *Heterodon simus* (southern hognose snake)
   30 Underside of tail noticeably lighter in color than belly; may be patterned above or may be black; length often exceeding 50 cm .. .. *Heterodon platyrhinos* (eastern hognose snake)
31 Scales along center of back weakly keeled or smooth; scales feel smooth when rubbed transversely 32
31 Scales strongly keeled; scales are noticeably keeled when rubbed transversely 33
   32 Basic pattern of blotching primarily orange and red; belly with black-and-white checkerboard pattern; top of head with V-shaped mark (figure 17) .... *Elaphe guttata* (corn snake)

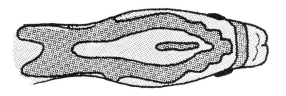

Figure 17. V-shaped mark on head of *Elaphe guttata*.

32 Basic pattern of blotching primarily gray or brown; top of head lacking V-shaped mark .... *Elaphe obsoleta* (rat snake)
33 Scale rows 27–33; belly plain white or yellow or with brown markings 34
33 Scale rows 23–25; belly solid red or with extensive brown and reddish markings 35
    34 Basic color light brown with darker brown or black squares on the back and side; belly with brown markings .... *Nerodia taxispilota* (brown water snake)
    34 Basic color dark gray to light olive above; sometimes with lighter mottling but no distinct pattern; belly gray or yellowish and unmarked except under tail .... *Nerodia cyclopion* (green water snake)
35 Uniformly black or dark gray above in adults; dark blotches apparent on back and sides of juveniles (less than 45 cm in length); belly usually solid orange or pink without conspicuous markings .... *Nerodia erythrogaster* (red-bellied water snake)
35 Series of reddish and brown crossbands above; belly yellowish with reddish or brown markings 36
    36 Mid-dorsal blotches connecting with lateral blotches throughout body length, making bands; some specimens may be nearly black dorsally, with brick red lateral blotches; ventral pattern of dark, unpaired, irregular blotches, or wavy lines ....*Nerodia fasciata* (banded water snake)
    36 Mid-dorsal blotches alternating with lateral blotches on posterior portion of body; ventral pattern of dark, paired half-moons .... *Nerodia sipedon* (northern water snake)

Species Accounts

The following accounts briefly describe pertinent aspects of the ecology of each species on the SRS and present information that might be helpful to an investigator or interested observer unfamiliar with the herpetofauna in this area. Our impression of the ecological status of each species is based on our own experiences and those of colleagues who have conducted research on the SRS. Specific locations are indicated on map 4. Information on the following is provided for each species when available: localities and habitats on the SRS, seasonal and daily activity patterns, reproductive patterns and population characteristics, food and feeding, body size, collecting techniques, general observations and comments, and pertinent references. Because the information includes only data from SRS findings, not general accounts, some categories may be absent for some species. The numbers at the end of each species account refer to SREL reprint numbers of publications in which reference is made to the species. The species within each of the major groups are in alphabetical order.

Salamanders

Ambystoma maculatum (spotted salamander)
Migrating adults and egg masses ($N=30$) of this species were found in January and February 1984 and in March 1987 in backwater pools along sloughs and streams on Risher Pond Road and along Steel Creek. All of these sloughs and streams, as well as Risher Pond, contain predatory fish. Spotted salamanders have never been collected at upland temporary ponds commonly inhabited by other species of *Ambystoma*.

Egg masses have an apparently symbiotic algae associated with them. Juveniles metamorphose in May and June at mean body sizes of 34.9 mm SVL. Joe Pechmann of SREL collected newly metamorphosed and larval *A. maculatum* on 5 June 1988 at Little Robbins Bay, near Robbins Station (approximately 3 km SE of Risher Pond). A few specimens have been captured at drift fences and Risher Pond. Most specimens have been found at night in shallow pools.

Until the winter of 1983–84, the spotted salamander was known only from five adults collected at Risher Pond. This species is considered uncommon on the SRS and probably is restricted to bottomland hardwood forests along streams and the Savannah River floodplain. SRS references 643, 716, 844, 863, 1162, 1244

Ambystoma opacum (marbled salamander)
The marbled salamander breeds in most temporary ponds and Carolina bays on the SRS and is consistently abundant at Ginger's Bay (located on the SRS between Jackson Barricade and city limits of Jackson). This species mates and lays eggs in autumn. Adults begin their migrations to breeding sites as early as August, but peak immigration into the bays occurs in October. Males emigrate after breeding. Females emigrate when the ponds fill. Larvae remain in aquatic habitats until April, May, or June, when they metamorphose and move onto land.

Breeding population sizes range from 20 to 10,000 adults. Females deposit 30–200 eggs in a small depression under logs, grass clumps, and debris in the basin of a dry pond. Eggs are tended by the female and hatching is delayed until the eggs are covered by water. *Ambystoma opacum* larvae are generally the first salamanders to hatch each year and consequently are usually the largest larvae found in ponds during early spring. Most larvae metamorphose in April and May at body sizes of 25–54 mm SVL. The largest metamorphs mature and return to reproduce at one year of age.

Adults can be seen on SRS highways during rainy periods in autumn and winter and captured with drift fences and pitfall traps at most aquatic sites. Several hundred juveniles have been captured with pitfall traps around Risher Pond during summer. Occasional individuals are uncovered under debris along woods bordering the Savannah River swamp and other aquatic areas. David Scott of the SREL captured, marked, and released more than 9000 individuals during autumn 1986 at Ginger's Bay. SRS references 590, 643, 658, 804, 844, 1280

Ambystoma talpoideum (mole salamander)
The mole salamander breeds in all types of temporary aquatic habitats and some permanent ponds. This species is found in virtually all aquatic habitats on the SRS that do not contain fish other than *Gambusia* and at some sites where fish are present (e.g., Risher Pond). Mole salamanders can be found migrating to breeding sites from October to March. Migra-

tions are nocturnal and are positively correlated with temperature and rainfall. Variation in the timing in migrations is related to annual variation in meteorological conditions. Adults emigrate in late winter or spring after breeding. Emigration of recently metamorphosed individuals from breeding sites occurs primarily from May to September when they migrate to surrounding terrestrial habitats. They remain in underground burrow systems until the next breeding season. Many terrestrial adults remain within 300 m of the breeding site and come back to the same pond each year.

Breeding population sizes can vary among sites and years from 100 to 6000 individuals, often with a sex ratio of 1:1. Females deposit from 10 to 1000 eggs singly and scatter them haphazardly across the pond bottom on leaves, grass, and twigs. Eggs hatch in 30–40 days and larvae are first found in February or March. Most larvae metamorphose in June and July, but the timing of metamorphosis is correlated with the time of pond drying. The mole salamander is facultatively paedomorphic in fish-free permanent ponds (e.g., Ellenton Bay, Flamingo Bay). Both paedomorphs and terrestrial morphs mature and can reproduce at one year of age and may live longer than nine years.

Larvae exhibit diel vertical migrations and feed mainly on zooplankton in the water column and aquatic insects and tadpoles in the benthos. Body size is variable among populations (adult SVL mean = 46.8–62.4 mm; metamorphosing juvenile mean = 43.5–49.1 mm).

Mole salamanders have been caught in enormous numbers with drift fences and pitfall traps (14,000 in one year at Rainbow Bay; 8000 in one day at Ellenton Bay). Aquatic individuals can be readily captured in minnow traps or by dip netting. This is one of the most abundant and best-studied amphibians on the SRS. SRS references 276, 590, 643, 658, 675, 684, 716, 743, 804, 844, 972, 1120, 1162, 1164, 1205, 1244, 1280, 1301

Ambystoma tigrinum (tiger salamander)
The tiger salamander breeds in many of the fish-free temporary and permanent ponds and Carolina bays throughout the SRS. Terrestrial adults are fossorial and remain in burrows in the surrounding areas until the next breeding season. Tiger salamanders are capable of digging their own burrows.

The tiger salamander breeds during winter from December through February. Emigration of adults from the breeding sites occurs in February or March. Migrations are nocturnal and related to temperature and rainfall. The actual timing of courtship and oviposition varies from year to year. Females deposit eggs in small clusters (24–98 eggs) and attach them to small twigs or plant stems in deep water. Eggs hatch in 30–40 days, and larvae metamorphose in May through July. Most metamorphs probably take two years to mature, but some males can reproduce at one year of age. Breeding population sizes vary from 3 to 180 individuals with a male-biased sex ratio of 1.2:1–1.9:1. Body sizes of adults and juveniles are as follows: male SVL mean = 94.1 ±0.38 mm; female mean = 95.5 ±0.53 mm; metamorphosing juvenile mean = 74.8 ±0.24 mm and 85.1 ±0.43 mm. Drift fences, dip netting for larvae, and searching for the distinct egg masses are all effective at revealing the presence of *A. tigrinum*.

This species was proposed as an endangered or threatened species at the South Carolina Endangered Species Conference in 1975. However, it is present seasonally at most temporary pond habitats on the SRS that have been sampled. The species was removed from consideration as one of special concern by the South Carolina Herpetological Taxa Review Committee in 1984. SRS references 276, 590, 643, 716, 804, 844, 863, 964

Amphiuma means (two-toed amphiuma)
Specimens have been collected in a variety of aquatic habitats, including Steel Creek Bay, Crystal Creek, and McQueen Branch; large numbers have been collected in the North Cove area of Par Pond with minnow traps during winter. Larger individuals have also been captured with small-mesh turtle (hoop) traps in 1984–85 below Risher Pond (N = 11) in February, April, May, and July; in sloughs alongside Steel Creek (N = 2) in November; at Dry Bay (N = 2) in July; and at North Cove (N = 10) in July, November, and February. Occasionally individuals are caught in terrestrial pitfall traps, indicating a tendency to move overland in some situations. SRS references 643, 840

Desmognathus auriculatus (southern dusky salamander)
Adults can be found under leaf litter in seepage areas alongside most or all small streams on the SRS. Larvae can be obtained by dip netting through leaf litter in small streams. This species is not particularly abundant anywhere on the site; it does not attain the population levels commonly observed in *Desmognathus* species in certain other parts of the range. Most specimens have been caught in damp areas of bottomland hardwoods beneath leaf litter. SRS references 643, 840, 848

Eurycea cirrigera (two-lined salamander)
These salamanders are occasionally found under litter alongside stream or swamp margins throughout the SRS. Many specimens have been collected on rainy nights on the section of Road F crossing the Upper Three Runs Creek floodplain. Larvae are commonly found in most small streams. Adults have been collected in drift fences along streams and larvae have

been taken with dip nets in leaf accumulations in small streams. SRS references 643, 677, 743, 804, 840, 848

Eurycea longicauda (long-tailed salamander)
Individuals are occasionally found under litter alongside stream or swamp margins throughout the SRS. Many specimens have been taken on rainy nights on the section of Road F crossing the Upper Three Runs Creek floodplain. Larvae are commonly found in most small streams on the SRS. Adults can be captured beneath logs and debris along streams and in swampy areas. SRS references 276, 643, 840, 848

Eurycea quadridigitata (dwarf salamander)
The dwarf salamander is a common pond-breeding, autumn-migrating species in most temporary ponds and Carolina bays. Adults can be found immigrating to temporary ponds from August to October. The initiation of breeding migrations appears to be associated with the seasonal decline in air temperature rather than rainfall. Migrations occur primarily at night but may be crepuscular.

Breeding population sizes can vary from 10 to 10,000 adults, with a 1:1 sex ratio. Oviposition does not occur until November or December, when the pond fills with water. Females deposit 7–48 eggs in a single cluster on vegetation in the water. Newly hatched larvae are first found in ponds in January and February at 6–7 mm SVL. After a five- to seven-month larval period, individuals metamorphose in June and July at 20-26 mm SVL. Adult males mature and reproduce at one year of age but females may delay reproduction until their second year. Adults feed on terrestrial insects and arachnids.

This species is captured commonly in pitfall traps along drift fences at temporary ponds on the SRS. Numerous individuals can be picked up during rainy nights on highways passing through low wooded areas. Two color morphs (yellow and brown) are common at two known localities (Bullfrog Pond, Sun Bay), but the brown morph predominates at other SRS sites. These may be different species as yet unnamed (Julian Harrison, pers. comm.). SRS references 643, 658, 675, 677, 706, 804, 1280

Necturus punctatus (dwarf waterdog)
Specimens have been found in tributaries of Upper Three Runs Creek and Meyers Branch. The species probably occurs in most other small streams on the SRS. A sample of 50 individuals was taken in several streams during October and November 1985 in sandy sections and in deep, muddy sections with leafy bottoms and undercut banks. Of 50 individuals dissected, 17

females and 19 males were mature adults (≥75 mm), and 14 individuals were juveniles (<75 mm). In 24 dissected individuals, dominant food items were earthworms (11%), insects (13%), salamanders (2%), crayfish (2%), and plant material (2%). SVL ranged from 26 to 116 mm (N = 50). Effective capture methods are unbaited minnow traps, electroshocking, and dip netting in areas with heavy leaf litter. SRS references 643, 1188

Notophthalmus viridescens (eastern newt; red-spotted newt)
This species is ubiquitous on the SRS. It is one of the few amphibians successful in permanent ponds and Carolina bays with predatory fish as well as in ephemeral ponds. Adults can be found, at least in small numbers, in almost all lentic situations, particularly Carolina bays. Flamingo Bay contains the largest concentration of newts yet found on the SRS. Terrestrial stages (efts) are frequently encountered on highways during rainy periods in late autumn, winter, and spring. The eastern newt is a prolonged breeder, starting in February or March and continuing through April or May. Terrestrial adults migrate to temporary ponds during the fall and winter. Adults often do not leave more permanent ponds unless they dry. Migrations by efts and adults are diurnal as well as nocturnal, presumably because of their aposematic coloration and distastefulness.

Breeding population sizes are large, ranging from 200 to 3000 adults. Females deposit eggs singly, hiding and wrapping each egg in vegetation. Juveniles metamorphose from June through September; 19,284 juveniles emigrated from Rainbow Bay in 1984. Paedomorphic newts have been found on the coastal plain of North Carolina and at two bays on the SRS. The terrestrial eft stage lasts for an unknown period of time on the SRS but is probably one to three years. Drift fences are effective for capturing terrestrial stages. Minnow traps can be used to capture aquatic forms. SRS references 643, 658, 804, 848, 1280

Plethodon glutinosus (slimy salamander)
This species is generally associated with woodland habitats (primarily hardwood, but also pine) and can be found commonly beneath litter throughout the SRS.

This is the only salamander on the SRS that does not have aquatic larvae. The adults presumably lay eggs underground or beneath litter in autumn, and the young of the year appear aboveground during January and February. Large numbers of males with prominent mental glands have been collected in terrestrial drift fences along Pickerel Pond during September and October.

Hand collecting and drift fences in hardwood habitats are the most effective techniques for capturing this ubiquitous terrestrial species. *Pleth-*

odon websteri is found a few kilometers north of the SRS but has not been found on the site. SRS references 643, 658, 666, 804, 848

Pseudotriton montanus (mud salamander)
Specimens are encountered occasionally on highways during rainy periods in late autumn and winter and are associated primarily with deciduous forests bordering streams. Larvae can be found in many of the small streams and seepage areas on the SRS and can be readily collected in leaf packs in streams with a dip net. This species is much less common than *P. ruber*. SRS references 643, 804, 840

Pseudotriton ruber (red salamander)
Red salamanders are common year-round in small streams and seepage areas or crossing roads on rainy nights. Females produce 80–120 large, yolky eggs during October and November. Adult females average 70.4 mm SVL. Hatchlings can be collected in most springs and small streams in January (mean = 11–14 mm SVL). Most larvae metamorphose from June to September at about 18–23 months of age (mean = 45–50 mm SVL). Larvae can be collected with unbaited minnow traps and by dip netting in leaf packs. This species is much more common than *P. montanus*. SRS references 643, 804, 840

Siren intermedia (lesser siren)
A concentration of this species was observed during a period of flooding along the shoreline of an inlet in Par Pond. Many captures have been made during autumn and winter in the North Cove area of Par Pond with minnow traps. This species has also been reported from a variety of other locations, including Crystal Creek, McQueen Branch, Rainbow Bay, and Bullfrog Pond. Specimens have been collected at Rainbow Bay, indicating the ability of this species to inhabit temporary bodies of water that predictably dry up. An individual was unearthed at Bullfrog Pond more than 20 months after the pond dried (David Scott, pers. comm.). More than 1000 *Siren intermedia* juveniles were captured in the vegetation and mud along the receding shoreline of Steed Pond during autumn 1981. This species probably occurs in most aquatic habitats on the SRS. Individuals (N = 40) were collected in 1984–85 in each month from January to May and in July, September, and October with minnow traps in the outlet stream below Risher Pond.

Monthly sampling indicates that some *S. intermedia* oviposit in March, but gravid females have been found in April. The smallest larvae have been collected in May and June at 10–12 mm SVL. Frequency distributions of body size during the summer indicate three distinct size classes, suggesting larvae take two years to reach adult body size and, presumably, sexual maturity. Baited or unbaited minnow traps and turtle traps are effective collecting techniques. It is noteworthy that *Siren intermedia* and *Siren lacertina* have been found to coexist only in the drainage below Risher Pond dam. SRS references 643, 840

Siren lacertina (greater siren)
Individuals were collected in 1984-85 in turtle hoop traps alongside Steel Creek at Road A (N = 22) in July, October, and November, and in Dry Bay (N = 29) in February and July. Several others have been collected in turtle traps in Dry Bay; a few individuals emerged from the water adjoining Steel Creek at Road A after a rotenoning operation. A large population is present at Dry Bay. Minnow traps and turtle traps are effective capture techniques. SRS reference 643

Frogs and Toads

Acris crepitans (northern cricket frog)
Infrequently reported but may be much more common than observed. Little is known of its ecology on the SRS. This species was heard calling in the Steel Creek delta on 16 May 1985 by Trip Lamb and Paul Moler. SRS references 222, 643

Acris gryllus (southern cricket frog)
Acris gryllus can be heard calling from almost all lake margins or Carolina bays on the SRS from spring to late summer. This species is abundant around many of the permanent ponds inhabited by fish (e.g., Dick's Pond). Color polymorphism occurs in the species, including at least green and brownish on the SRS. The tadpoles are often found in aquatic habitats having predatory fish. Most specimens have been collected by hand or in pitfall traps along drift fences. Most SRS specimens of *Acris* have been *A. gryllus*, although some *A. crepitans* also occur on the site. SRS references 222, 276, 279, 280, 362, 643, 804

Bufo quercicus (oak toad)
Individuals of this species are encountered less frequently than *B. terrestris*. A few choruses have been heard at Pump Station #1 and in a depression at the junction of Road 2 and Road F in early summer. Despite the lack of heavy concentrations, this species appears to occur throughout the SRS but is apparent in large numbers only after heavy rainfalls when choruses

emerge in low-lying temporarily flooded areas. Drift fences capture some individuals, but the largest numbers are found by locating breeding choruses. SRS references 222, 279, 280, 362, 643, 658, 804

Bufo terrestris (southern toad)
This is the most conspicuous amphibian on the SRS throughout the warm periods of the year. Specimens can be seen on highways on most warm nights. Individuals are captured on a regular basis at all drift fence sites, including upland pine areas, where the species is abundant.

Breeding usually begins in pools of standing water in early spring (late March or early April) after rainfall. The eggs are laid in long strings. In some years (e.g., 1983) few *B. terrestris* breed on the SRS, presumably because of the lack of rainfall during April and May. Large numbers can be collected by drift fences, road collecting, or locating calling adult males during breeding periods. SRS references 222, 276, 280, 362, 392, 475, 643, 658, 664, 804

Gastrophryne carolinensis (eastern narrow-mouthed toad)
This species is ubiquitous on the SRS and has been captured at all drift fence localities. Small choruses of narrow-mouthed toads can often be heard in temporary pools of standing water in late spring or summer throughout the SRS. Joe Pechmann of the SREL reported a breeding congregation in September 1982 at Rainbow Bay. Individuals are frequently encountered under litter and on highways.

Large populations are usually associated with open grassy habitats rather than woodland sites. Females deposit a floating monolayer of eggs on the surface of the water in thick grassy vegetation. Drift fences capture large numbers of this species at some locations. SRS references 222, 280, 362, 643, 658, 804, 1301

Hyla avivoca (bird-voiced treefrog)
The greatest concentrations of this species on the SRS have been observed at locations bordering the Savannah River swamp, particularly in association with cypress trees. Large choruses have been heard behind the Hog Barn, below Risher Pond, and along the margins of Steel Creek, particularly near the delta area. Individuals are occasionally captured in drift fences on the interior portions of the SRS, but these presumably represent smaller populations than those found in the swamp. This is a summer-breeding species. Searching at sites where individuals are calling is the most effective collecting technique. SRS references 222, 643

Hyla chrysoscelis (Cope's gray treefrog)

Primarily a late spring and summer breeder (May–August), individuals might be encountered anywhere on the SRS. This species is particularly common in the low-lying areas near the swamp. Large choruses have been reported from Water Gap Road during the summer. Small populations occur at a variety of sites, including Rainbow Bay, Risher Pond, and Morse Code Bay. Males call from the edges of ponds or from nearby vegetation. Females deposit multiple small floating clusters (5–20) of eggs on the surface of small pools or ponds. The most effective technique for locating individuals is to listen for breeding choruses.

This species is frequently confused with the sibling species *Hyla versicolor* but can be distinguished from the latter by its faster trill rate. Also, *H. chrysoscelis* is diploid (2N) whereas *H. versicolor* is tetraploid (4N). J. T. Collins (pers. comm.) did chromosome counts on two specimens of gray treefrogs collected on the SRS and determined them to be diploid. SRS references 222, 643, 804, 1222

Hyla cinerea (green treefrog)

Large choruses are found around Par Pond and throughout the SRS in farm ponds and Carolina bays, and along the river swamp or deltas during late spring and summer. Choruses are less dependent on rain than most other species; green treefrogs frequently call at night from emergent vegetation even after several days or weeks without rain. The most effective capture technique is hand collecting. SRS references 222, 276, 362, 429, 643, 658, 804

Hyla femoralis (pine woods treefrog)

Small numbers of calling individuals have been noted at a variety of sites throughout the SRS. Large choruses have been reported at Rainbow Bay. This is a summer-breeding species. Males call from trees or shrubs surrounding temporary ponds, and females deposit small floating clusters of eggs. Most individuals have been collected at night by locating breeding choruses, and some have been captured in pitfall traps. SRS references 222, 262, 276, 362, 643, 804, 1301

Hyla gratiosa (barking treefrog)

This is a locally common species, and specimens are occasionally found on highways during rainy periods in the spring. Choruses have been located along the river swamp and at Karen's Pond, Rainbow Bay, and Flamingo Bay. Males typically call while floating on the water's surface during spring and summer breeding. Most individuals are collected at night by locating breeding choruses. SRS references 222, 280, 643, 804

Hyla squirella (squirrel treefrog)
Although enormous choruses of this ubiquitous species are common on coastal islands near Charleston, S.C., they are seldom heard in large numbers on the SRS. Individuals are frequently encountered on highways at night or during rainy periods and are occasionally heard calling singly during the day at a variety of locations on the SRS. Occasionally collected at drift fences or by hand. SRS references 222, 276, 280, 362, 643, 658, 684, 804

Pseudacris crucifer (spring peeper)
This species, known until recently as *Hyla crucifer*, can be heard calling from most aquatic areas from November to March and occasionally at other times. It is ubiquitous on the SRS but is especially abundant in temporary ponds during the breeding season. This species is frequently captured at drift fences and by locating breeding choruses. SRS references 222, 262, 276, 280, 362, 643, 658, 1051

Pseudacris nigrita (southern chorus frog)
Pseudacris nigrita is primarily an upland species on the SRS and occurs in small numbers in most temporary ponds. Individuals move to breeding ponds as early as December but can breed as late as February or March.

Breeding population sizes can vary among sites and years from 4 to 129 individuals, with a sex ratio of 1:1. Eggs are deposited generally in February or March. Juveniles metamorphose between April and June at body sizes of 11.5–16.0 mm SVL, and sexual maturity may be reached at one year of age. Adult body sizes for males are 26.0–32.5 mm SVL, and females 25.0–33.0 mm SVL. Drift fences are effective for capturing large numbers. SRS references 222, 280, 643, 1051, 1118

Pseudacris ocularis (little grass frog)
This species, formerly known as *Limnaeodus ocularis* and *Hyla ocularis*, is found in scattered localities throughout the southern portion of the SRS. Individuals can frequently be heard during the warmer months along the river swamp margin. It has been found most commonly in small choruses in grassy areas along flooded roadside ditches. Hand collecting of calling individuals is occasionally effective; few have been captured by other means. SRS references 222, 362, 643, 804

Pseudacris ornata (ornate chorus frog)
Pseudacris ornata is primarily an upland species occurring in most temporary ponds and Carolina bays. It is two or three times more abundant than *P. nigrita* at sites studied in detail. It migrates to breeding sites as early as

November, but peaks of immigration do not occur until January or February. Emigration of adults occurs in January, February, and March.

Breeding population sizes have been found to vary among sites and years from 40 to 600 individuals, with a sex ratio of 1:1. Eggs are deposited in January, February, or March in small masses attached to grasses or twigs near the surface of the water. Juveniles metamorphose between April and June at body sizes of 18.0–22.5 mm SVL. Sexual maturity can be reached at one year of age. This species displays a color polymorphism (green, brown or copper, and gray), with the green morph the least common.

Adult body sizes of males range from 31.0 to 39.0 mm SVL, and females from 33.0 to 40.0 mm SVL. Drift fences are the most effective technique for capturing adults and recently metamorphosed individuals. SRS references 222, 280, 362, 643, 804, 1051, 1118, 1301

Rana areolata (crawfish frog, Carolina gopher frog)
Crawfish frogs have been heard calling from Karen's Pond, and several individuals were captured with pitfall traps in 1969–70. A few were similarly taken at Risher Pond, Sun Bay, Ellenton Bay, and Flamingo Bay. This species is probably widespread in association with aquatic habitats on the SRS, but seldom have more than 10 individuals been found at any single locality. Adults have been heard calling at many of the Carolina bays.

Adults were captured in pitfall traps on 16 February 1982, and metamorphosing juveniles were collected on 31 May 1982 at Flamingo Bay. Justin Congdon observed a male and female in copulation underwater at a Carolina bay near Dry Bay on 6 March 1989. Most specimens have been captured in pitfall traps along drift fences. SRS references 222, 362, 643, 804

Rana catesbeiana (bullfrog)
Bullfrogs are found in virtually every permanent body of water on the SRS but are seldom, if ever, the most abundant species present. Juveniles have been captured in pitfall traps several meters away from water. The bullfrog becomes active at breeding sites in spring, and calling begins in April or May and extends through the summer.

This large frog may breed in moderate-sized temporary ponds with population sizes of fewer than 10 individuals. However, the successful metamorphosis of juveniles from these ponds occurs only in years when the ponds do not dry until late summer (e.g., Rainbow Bay). Bullfrogs are most abundant in permanent farm ponds, large Carolina bays, and reservoirs on the SRS (e.g., Par Pond, Steed Pond, Ellenton Bay). Breeding populations at the latter sites can exceed 20 or 30 individuals. The female deposits a large floating film of eggs (5000–20,000) in thick vegetation of shallow-water littoral zone habitats. Eggs can be found from May to July. Tadpoles are capable of metamorphosing in 4–5 months on the SRS but may overwinter in permanent ponds and metamorphose the following April or May at 12–13 months of age. Bullfrogs are among the frog species whose tadpoles can live in ponds inhabited by predatory fish.

Adults are most easily collected by hand or dip net around lake margins at night. Numerous smaller individuals and some adults have been captured in pitfall traps. SRS references 209, 222, 272, 279, 362, 643, 658, 804

Rana clamitans (bronze frog)

Bronze frogs occur in both temporary and permanent ponds throughout the SRS, and egg masses and tadpoles are commonly found in small streams. Calling individuals can be heard during most of the warm months of the summer. The species is clearly capable of coexisting with predatory fish. Drift fences with pitfall traps are an effective means of capture. SRS references 362, 643, 658, 804

Rana grylio (pig frog)

A small chorus was heard calling from Steel Creek Bay twice during the spring of 1977. The species was also reported from the SRS by Freeman (1956). R. Humphries of SREL reported this species from the river swamp on the SRS in 1953–54. Individuals have been heard calling from Steel Creek Bay, and a single individual was heard and seen calling at North Cove in Par Pond. Individuals are occasionally seen in the Savannah River swamp but large choruses are uncommon. Because of its scarcity, little is known about the ecology of the species on the SRS. Hand collecting is presumably the most effective approach for capturing pig frogs. SRS references 222, 276, 280, 643

Rana palustris (pickerel frog)

A few specimens were captured in pitfall traps in 1968–69 at Risher Pond. Individuals were collected in 1981 and 1986–87 at Ellenton Bay and in the delta of Steel Creek. No large breeding choruses of this species have been reported from the SRS, and they are apparently uncommon in the few habitats where they have been found here. Most specimens have been captured by pitfall traps along drift fences. SRS references 222, 362, 643

Rana sphenocephala (southern leopard frog)

Numerous individuals can be collected on SRS highways after winter rains, often long distances from water. They can be collected from all aquatic areas on the SRS. Large numbers of breeding adults have been captured with pitfall traps and terrestrial drift fences during winter at Ellenton Bay and Risher Pond.

This species has been reported to have two peaks of reproduction, one in January–March and the other in September–October. Tadpoles that hatch in the fall overwinter in the pond (Joe Pechmann, pers. comm.). Egg masses oviposited in the spring have been reported to be laid communally, whereas those in autumn are isolated (Jan Caldwell, pers. comm.).

Large numbers can be captured with drift fences. In 1982 the authors removed more than 600,000 recently metamorphosed leopard frogs from pitfall traps on the inside of the drift fence at Ellenton Bay within a one-month period. The unusual abundance was attributed to a major drought the previous year that reduced water levels and resulted in a flush of nutrients and the elimination of many aquatic predators, resulting in a higher survival rate of the tadpoles. SRS references 222, 276, 280, 362, 643, 658, 804, 863, 1008, 1051

Rana virgatipes (carpenter frog)
A small chorus of a dozen individuals was heard and four were collected at Steel Creek Bay in early summer 1977. Choruses were also heard in 1979–80 at Steel Creek Bay. The species has been discovered at a variety of sites along Steel Creek and in the delta. A single specimen was observed by Jeff Lovich on 24 May 1987 at Ellenton Bay. The few individuals captured have been collected by hand. Following the report of Humphries (1953–54), who observed or heard the species at a locality presumed to be Steel Creek Bay, carpenter frogs were not verified again by capture on the SRS until 1977. SRS references 222, 643

Scaphiopus holbrooki (eastern spadefoot toad)
Heavy rains appear to be the stimulus for breeding choruses of this species, and many habitats on the SRS that flood temporarily serve as breeding sites. Large choruses are commonly found in man-made habitats (e.g., borrow pits). Metamorphs or juveniles have been captured by drift fences throughout the year.

In three years, reproduction occurred in late February or during March (23 March 1979, 11 March 1980, 28 February 1987) following brief, heavy rains (50 mm in less than 24 hours). Large choruses were heard after heavy rains in June 1989 (Joe Pechmann, pers. comm.). In some years, no breeding choruses have been observed on the SRS.

Drift fences and road collecting are effective capture techniques during rainy periods. Breeding populations can be located by listening for calling adult males, which are audible for up to 1.5 km, after heavy rains. SRS references 222, 276, 280, 362, 643, 658, 792, 804

Alligators

Alligator mississippiensis (American alligator)
Breeding adults are present on the site, particularly in the Par Pond and Beaver Dam Creek systems. Nests have been found at Upper Three Runs Creek, Par Pond, Pond C, Steed Pond, and Beaver Dam Creek. Several successful hatches have been observed in the Par Pond system.

The approximate number of adult alligators on the SRS is 200–250 (estimated by Rich Seigel). Most are concentrated in Par Pond, but as many as 40 are estimated to occur at Steel Creek, and a minimum of 28 adults and 20–30 hatchlings were observed at Beaver Dam Creek during aerial surveys. Much smaller populations (fewer than 10 individuals) are found at Pond B, Pond C, Four Mile Creek, Upper Three Runs Creek, and Pen Branch. We know of two instances where alligators occurred in Carolina bays (Ellenton Bay, 1969; Flamingo Bay, 1977) for at least two years and then disappeared from the habitats and have not been seen since.

Small alligators (up to 1 m) can be captured by hand or dip net at night or occasionally in baited hoop nets. Larger individuals are more wary but can sometimes be captured with a wire cable noose at night. An effective technique for capturing large alligators is the use of trap boards as guides into a noose attached to a bent bamboo pole that is triggered when the bait (fish) is taken (Murphy and Fendley, 1975).

American alligators on the SRS were spared the heavy poaching pressure of the 1950s and 1960s because of the protection afforded by the site security system. This may explain the presence of relatively large numbers in a variety of SRS habitats since the mid-1960s. The largest individual (3.84 m) was captured in Par Pond by Laura Brandt in May 1988. SRS references 79, 209, 326, 391, 405, 440, 553, 566, 612, 643, 664, 701, 717, 744, 775, 812, 821, 1078, 1304

Turtles

Chelydra serpentina (snapping turtle)
Although large numbers are unlikely to be found at any one site, single specimens of this ubiquitous species may occur in or within several hundred meters of almost any aquatic habitat on the SRS. Snapping turtles

have been captured in aquatic and terrestrial habitats on the SRS during every month. However, a total of only five individuals have been caught in November, December, and January.

Individuals of *C. serpentina* have been marked in population studies at a variety of locations on the SRS: Ellenton Bay (N = 142), Risher Pond (N = 30), Steed Pond (N = 24). Clutch size varies from 17 to 29 (N = 5, mean = 25). The largest *C. serpentina* captured on the SRS weighed 10 kg. The mean size of adult males is 196 mm (PL); of females 186 (PL).

The most effective capture method for adult turtles is with baited aquatic traps, although single individuals are frequently encountered moving overland. Hatchlings and juveniles are frequently captured in pitfall traps. Adults are often captured alongside drift fences and occasionally in pitfall traps.

Noteworthy but unexplainable is the fact that few snapping turtles have been captured in Pond B (N = 1) or Par Pond (N = 14) in more than 17 years of trapping in these areas. SRS references 209, 326, 507, 565, 566, 643, 678, 742, 804, 827, 852, 892, 908, 999, 1007, 1019, 1073, 1127, 1342

Chrysemys picta (painted turtle)
The maximum weight of the 10 males caught on the SRS was 0.221 kg. The mean body size was 112 mm (PL). The few captured on the SRS have been at drift fences or crossing roads. Where this species is abundant, individuals are readily captured in baited hoop traps.

The painted turtle was not recorded from the SRS until 1976, when the first individual was captured in a pitfall trap at Ellenton Bay. Subsequently, a total of 10 individuals has been captured on the site; all have been males. This suggests that the SRS is on the margin of the painted turtle's geographic range and that populations occur in the region. Those found on the SRS may be transient males that have traveled long distances from their usual populations. Populations exist immediately northwest of the SRS near Jackson and Ridge Spring, S.C. SRS references 199, 200, 216, 263, 507, 565, 630, 643, 678, 718, 742, 793, 812, 816, 827, 852, 858, 868, 1007, 1151, 1342

Clemmys guttata (spotted turtle)
Only one relatively large concentration of this species has been found on the SRS. Between 4 March and 26 April 1988, 10 spotted turtles were captured at a drift fence as they entered Squirrel Bay. Two others entered at nearby Ginger's Bay. Other spotted turtles have been picked up on roads at the following locations: Risher Pond Road, S.C. 125 (SRP Road A) near Upper Three Runs, Morse Code Bay, Road A near Steel Creek Bay, and

Road 3 near Bulldog Bay. Presumably populations are somewhere in the vicinity of each location. Most specimens have been collected in early spring (March). Tony Mills and Jeff Lovich collected three specimens in the seepage area, bog-type habitats alongside Risher Pond Road on 1 March 1987.

The largest *Clemmys* caught on the SRS weighed 183 g. The mean plastron length is 86 mm. Most of the spotted turtles captured on the SRS were picked up on highways. Jeff Lovich observed that males of the Risher Pond Road population repeatedly make extensive overland movements between wetland habitats in the spring (up to 423 m in 24 hours). Hand capture of individuals in early spring in certain areas appears an effective capture technique. Drift fences are sporadically effective. SRS references 565, 566, 643, 1282

Deirochelys reticularia (chicken turtle)
Chicken turtles occur most commonly in Carolina bay habitats but are found in small numbers in other aquatic areas. A unique feature of this species (among SRS turtles) is the reproductive cycle of the females. Females on the SRS lay eggs from autumn through winter and early spring rather than in late spring and summer. Clutch size varies from 2 to 14 eggs (N = 113, mean = 7.3). The largest *D. reticularia* caught on the SRS had a plastron length of 194 and weighed 1344 g. The mean size of adult males is 99 mm (PL); mean size of females is 161 mm (PL).

Males can be trapped effectively in aquatic areas, but females are captured most frequently in terrestrial drift fences and pitfall traps as they move to nesting sites. SRS references 200, 209, 244, 263, 326, 565, 566, 570, 603, 643, 652, 779, 804, 816, 852, 858, 866, 999, 1007, 1019, 1127, 1151

Kinosternon bauri (striped mud turtle)
This species has been reported from Steel Creek delta, Steel Creek, Ellenton Bay, and Pen Branch delta.

The largest male was 96 mm (CL), while the largest female was 112 mm. The mean CL of males is 85 mm (N = 9), and of females, 98 mm (N = 13). Most specimens have been collected with baited hoop traps or by hand while they were traveling overland.

This species has had one of the most interesting histories of any reptile on the SRS. Its presence was not reported by Freeman (1955b), but Duever (1972) indicated that he had collected four specimens at Steel Creek. Gibbons and Patterson (1978) subsequently contested the records because three of the specimens were identified by experts as *K. subrubrum*. The identity of the fourth was questionable. However, the problem was solved by Trip Lamb, who used discriminant analysis to examine a series of *Kinosternon* from the Steel Creek area (1983a, 1983b). He concluded that both species of *Kinosternon* exist on the SRS, with *K. bauri* being found primarily in the Steel Creek area and along the margins of the Savannah River swamp. The SRS specimens lack the distinctly striped carapace characteristic of this species in other parts of the range. The yellow head stripes, though prominent in juveniles, become obliterated with age in some individuals, but in nearly all they remain visible between eyes and snout. Except for the SRS animals, only one *K. bauri* had been reported from South Carolina (collected in Jasper County, about 80 miles south of Steel Creek; Robert Shoop, pers. comm.) prior to 1986. Since 1987, the species has been confirmed to be present at the Webb Wildlife Center (Hampton County), in Bamberg County, and in Clarendon County. SRS references 643, 875, 882

Kinosternon subrubrum (eastern mud turtle)
This species is characteristically associated with bodies of standing water, particularly those with fluctuating levels such as Carolina bays and cypress-gum swamps. Specimens have not been reported from Par Pond, the streams, or the river. Mud turtles have been caught in all months of the year on the SRS.

Individuals of *K. subrubrum* have been marked in population studies at a variety of locations on the SRS: Ellenton Bay (N = 1356), Risher Pond (N = 158), Rainbow Bay (N = 297), Karen's Pond (N = 33), Flamingo Bay (N = 59), Steel Creek (N = 59). Clutch size varies from two to five (N = 223, mean = 3.1). The largest *K. subrubrum* captured on the SRS weighed 236 g. The mean size of adult males is 87 mm (CL); of females, 88 mm (CL).

Aquatic trapping frequently yields specimens, and terrestrial drift fences and pitfall traps are extremely effective when situated along the margins of Carolina bays. Many captures are made terrestrially because individuals hibernate on land and may move long distances from water or away from temporary aquatic habitats that are in the process of drying. SRS references 209, 244, 263, 318, 326, 565, 566, 603, 643, 652, 742, 779, 804, 816, 827, 866, 868, 875, 882, 999, 1007, 1019, 1073, 1127

Pseudemys concinna (river cooter)
This is the most conspicuous turtle along the Savannah River, where it basks on logs and exposed rocks in large numbers during the warmer months. A few individuals have been collected on the SRS, including Steel Creek delta (N = 10) and Four Mile Creek (N = 6) during reactor shutdown. Two females with clutches of 21 eggs (determined by X-ray)

have been caught. The largest *P. concinna* caught on the SRS was a female that weighed 5.4 kg and was 317 mm (PL). Sporadic captures have been made using large hoop nets or electroshockers in the Savannah River swamp. The species is seldom captured in baited aquatic traps, which are effective for other turtles. SRS references 507, 566, 643, 836, 999

Pseudemys floridana (Florida cooter)

This species occurs in most large aquatic habitats, including Carolina bays, streams, Par Pond, farm ponds, and the river swamp, but is seldom found in large numbers.

Individuals have been marked in population studies at a variety of locations on the SRS: Ellenton Bay (N = 216), Risher Pond (N = 71), Steed Pond (N = 31), Par Pond (N = 26), Pond B (N = 14). Clutch size varies from 4 to 24 eggs (N = 31, mean = 10.0). The largest *P. floridana* captured on the SRS weighed 3.9 kg. The mean size of adult males is 163 mm (PL); of females, 231 mm (PL). Although *P. floridana* occurs in habitats with elevated aquatic temperatures, this species is not known to achieve the rapid growth and increased size that *Trachemys scripta* does in these areas. Aquatic traps and pitfall traps have yielded the most specimens, but no one method has been highly effective. SRS references 209, 326, 520, 565, 566, 643, 779, 816, 866, 892, 999, 1007, 1019, 1127

Sternotherus odoratus (stinkpot)

Present in most permanent bodies of water on the site, stinkpots have been captured every month of the year on the SRS. Individuals have been marked in population studies at a variety of locations on the SRS: Ellenton Bay (N = 241), Risher Pond (N = 108), Steed Pond (N = 68), Steel Creek (N = 88), Par Pond (N = 70), Dick's Pond (N = 14), and Pond B (N = 12). Clutch size varies from two to eight eggs (N = 41, mean = 5.5). The largest *S. odoratus* captured on the SRS weighed 0.28 kg. The mean size of adult males is 91 mm (CL), and 92 mm (CL) for females. The most effective capture method is with baited aquatic traps. This highly aquatic species is seldom encountered terrestrially. SRS references 209, 244, 263, 318, 326, 565, 566, 603, 643, 652, 742, 779, 804, 816, 827, 866, 868, 875, 999, 1007, 1019, 1073, 1127, 1184

Terrapene carolina (box turtle)

Box turtles appear to be ubiquitous on the SRS but are seldom encountered except as solitary individuals. Many captures are on highways during the morning, especially on sunny days after a rain. No large populations of box turtles have been found on the SRS, although large numbers have been reported in hardwood habitats in other parts of Aiken County approximately 20 km north of the SRS (Michael Gibbons, pers. comm.). The upland hardwood habitats where this species appears to be common in other areas are limited on the SRS. The northern perimeter of the site would be the area most likely to support significant numbers of this species. Clutch size varied from two to four (mean = 3) in five gravid females captured on the SRS. SRS references 302, 565, 566, 643, 804, 999, 1073, 1111

Trachemys scripta (slider turtle)
This is the most frequently encountered turtle on the SRS and usually the dominant species in any lentic habitat that retains water year-round. Specimens have been found at practically every aquatic site.

Individuals of *T. scripta* have been marked in population studies in a variety of locations on the SRS: Ellenton Bay (N = 1568), Risher Pond (N = 240), Steed Pond (N = 375), Par Pond (N = 1037), Lost Lake system (N = 920), Ponds B and C (N = 477), Steel Creek (N = 399). Clutch size varies from 2 to 17 eggs (N = 232, mean = 7.7). The largest *T. scripta* captured on the SRS weighed 7.7 kg. A major body size discrepancy exists among adult individuals from different aquatic habitats on the SRS. Those from thermally altered areas such as Par Pond and some sections of the swamp are large (mean PL of males = 158 mm, N = 354; mean PL of females = 234, N = 760). Those from natural habitats are smaller (e.g., Ellenton Bay, mean PL of males = 139, N = 570; mean PL of females = 186, N = 353). The differing body sizes are a consequence of significantly faster growth rates of juvenile turtles in habitats with elevated water temperatures. Males become melanistic with increasing body size and age.

All means of trapping are effective, but baited aquatic traps and pitfall traps have yielded the largest numbers. SRS references 199, 200, 209, 216, 227, 244, 326, 381, 384, 405, 507, 565, 566, 612, 629, 630, 643, 664, 678, 682, 717, 718, 722, 742, 779, 793, 804, 816, 827, 836, 852, 858, 866, 868, 892, 908, 999, 1007, 1019, 1044, 1091, 1112, 1127, 1151, 1260, 1268, 1342

Trionyx spiniferus (spiny softshell turtle)
Several individuals have been seen and collected in Lower Three Runs Creek below the Par Pond outfall and in Steel Creek near Road A. Extensive trapping efforts in Par Pond and other lentic habitats on the SRS have failed to capture any, and they are presumed absent. The species apparently does not occur in any of the other lakes on the site. Several individuals have been trapped in the Savannah River below the SRS. SRS references 565, 566, 643

Lizards

Anolis carolinensis (green anole; chameleon)
This species is occasionally found in small populations where individuals can be seen regularly on suitable days during any month. Preferred habitats appear to be more mesic than those of *Sceloporus undulatus*, but it is apparently ubiquitous and active throughout the year on warm days. Large populations have been observed in the vicinity of the SREL's Aquatic Ecology Laboratory along Upper Three Runs Creek, Hog Barn, Flamingo Bay, behind the SREL Waterfowl Facility, and at Dry Bay. Hand collecting in the morning is an effective capture technique. SRS references 276, 315, 566, 643, 804, 869, 963, 1336

Cnemidophorus sexlineatus (six-lined racerunner)
This species can generally be found in any relatively open sandy area, including road shoulders and adjoining habitat. Individuals are frequently seen crossing highways during daylight hours in warm months. They normally do not become active until late April or May and begin hibernation in early fall (see Bellis, 1964). SRS references 48, 276, 566, 643, 676, 804

Eumeces fasciatus (five-lined skink)
Eumeces inexpectatus (southeastern five-lined skink)
Eumeces laticeps (broadheaded skink)
All three of these species of *Eumeces* have been documented as occurring on the SRS. *Eumeces fasciatus* appears to be restricted to moist areas with hardwood forest and is abundant in and around the margins of the Savannah River swamp. In contrast, *Eumeces inexpectatus* is widespread in relatively dry upland habitats having sandy soil, but does not appear to attain the high densities of *E. fasciatus*. The habitat preferences of *E. laticeps* are broader, allowing it to occupy most habitats used by the other two species; however, it is most abundant in lowlands with hardwood forest at the margins of the Savannah River swamp. All three species sometimes occur at the same site in the transition zone between swamp margin and upland forest.

These skinks emerge from hibernation in March on the SRS and become inactive during cold spells in autumn. By April the orange head coloration of males intensifies and the testes have become enlarged. Males are highly aggressive toward conspecifics from this time until June, when the head coloration fades and testis volume decreases (Vitt and Cooper, 1985). Females complete vitellogenesis in the second half of May, at which time they produce a pheromone that stimulates courtship (Cooper et al., 1986).

Mating takes place primarily in late May on the SRS. Females brood a single clutch of eggs through June until hatching in late July. Eggs are usually deposited under bark of fallen logs or standing dead trees. From emergence of hatchlings until the winter inactivity period begins, adults are solitary, engaging in little or no social interaction, including sexual or aggressive behavior (Vitt and Cooper, 1985). Most individuals are captured by hand. SRS references (*Eumeces fasciatus*) 566, 643, 804, 869, 886, 903, 963, 1011, 1022, 1050, 1056, 1095, 1102, 1110, 1169, 1326; (*Eumeces inexpectatus*) 566, 643, 804, 869, 903, 1022, 1050, 1056, 1095, 1110, 1169; (*Eumeces laticeps*) 566, 643, 804, 869, 886, 903, 963, 993, 1011, 1031, 1032, 1033, 1050, 1056, 1063, 1064, 1095, 1102, 1110, 1143, 1160, 1169, 1211, 1297, 1309, 1326

Ophisaurus attenuatus (slender glass lizard)
Ophisaurus ventralis (eastern glass lizard)
These lizards have been found in sandy habitats throughout the SRS but never in abundance. Otherwise, little is known of their ecology or distribution on the SRS. Most specimens have been captured in early morning. An SRS female *O. ventralis* laid five eggs and brooded them until they hatched (Laurie Vitt, pers. comm.). A female *O. ventralis* captured at Dry Bay by Tony Mills laid six eggs on 23 July 1989. The young hatched in early September. Drift fences in sandy areas have yielded a few individuals of each species. Occasional specimens are picked up on roads. SRS references 566, 643, 804

Sceloporus undulatus (eastern fence lizard)
Populations of fence lizards are generally restricted to disturbed areas, particularly where ground cover is available. Small concentrations can be found at most old homesites, brush piles, and bridges. Individuals can frequently be found in habitats of turkey oak and longleaf pine (see Tinkle and Ballinger, 1972). Most individuals are captured by hand. SRS references 276, 566, 643, 662, 676, 782, 804, 812, 1139

Scincella lateralis (ground skink)
Ground skinks apparently occur in all wooded habitats on the SRS or even in open areas if sufficient ground cover is available. They are active during warm, sunny periods in winter, as well as in other seasons. Drift fences successfully capture specimens, but the most effective collecting technique is the use of cover boards at appropriate sites. More than 50 per day have been captured at the SREL study area in a pine habitat on SRS Highway 1 at the junction with SRS Highway 1A North. SRS references 643, 804, 886, 903

Snakes

Carphophis amoenus (worm snake)
The worm snake has been collected at only one site on the SRS — an excavation pit dug during archaeological studies in the bottomland hardwood habitat along Pen Branch. The first specimen was reported by Jim Knight. All four specimens were collected by hand under litter. In other regions (Edgefield and McCormick counties) this species is readily captured by routine hand collecting, so it is presumed to be extremely rare on the SRS. SRS reference 566

Cemophora coccinea (scarlet snake)
This species is generally restricted to turkey oak–pine forests and abandoned old fields in sandy terrain but is found throughout the SRS. Specimens have been found at Risher Pond, Karen's Pond, Sun Bay, Ellenton Bay, Dry Bay, and Pond C. Scarlet snakes are almost strictly nocturnal and extremely seasonal in their activity patterns, being found only from April through September. Of the more than 200 individuals that have been captured on the SRS, none has been found in the other months of the year.

Our studies can contribute little to the sparse information available about the reproductive cycle of this species. However, a recently hatched juvenile was captured in May emerging from an egg that was in sandy soil 1.3 m below the surface. The mean SVL of females is 23 cm. Almost all individuals have been collected on SRS highways at night or in drift fences. SRS references 316, 566, 643, 804, 1133

Coluber constrictor (racer; black racer)
Racers can be found in any terrestrial habitat on the SRS, being particularly common in abandoned old fields, pine woods, and hardwood areas. This species is captured most frequently alongside drift fences or by hand. They are seen commonly from April to November, with some records from every month of the year. Mean SVL and body mass of males are 84 cm and 169 g, and of females, 83 cm and 150 g.

Black racers are the most frequently encountered terrestrial snakes on the SRS. They are often seen crossing highways, but during daylight hours only. SRS references 276, 301, 396, 397, 439, 566, 643, 804, 886, 903, 1133

Diadophis punctatus (ringneck snake)
Ringneck snakes are found primarily in deciduous woodland areas with heavy ground litter but are also found in pine-dominated habitats. Occasionally, specimens have been collected with drift fences around Pond C,

Risher Pond, Flamingo Bay, and Ellenton Bay. The largest known population is at Dry Bay. As many as six to eight have been collected at Dry Bay in a day on several occasions. They are infrequently encountered at most locations on the SRS but have been collected in every month of the year. The mean SVL of both sexes is about 21 cm. Most specimens have been captured at drift fences, and individuals are seldom seen crossing roads. SRS references 566, 643, 742, 804, 1133

Elaphe guttata (corn snake)
Corn snakes are seldom seen in swampy areas and are encountered most frequently in dry, terrestrial habitats. They are found throughout the SRS, generally associated with woodland habitats, including pine and hardwood areas. The corn snake is predominately active at night from April to September. They are not collected from January to March. Mean SVL and body mass of males are 89 cm and 224 g, and of females, 73 cm and 136 g. They are commonly collected at night on SRS highways or during the day in cooler periods, and also found at drift fences in most areas. SRS references 315, 396, 439, 475, 566, 624, 643

Elaphe obsoleta (rat snake)
Rat snakes are found in a wide variety of terrestrial or semiaquatic habitats throughout the SRS but are especially common in forested swampy areas. They are frequently found in wood duck nest boxes during spring and are known to eat the duck eggs. Rat snakes are usually diurnal on the SRS, although nocturnal activity is sometimes observed. Most individuals are captured between March and October. Mean SVL and body mass for males are 133 cm and 666 g, and for females, 131 cm and 559 g. Most specimens are captured by hand. SRS references 396, 397, 439, 566, 643, 710, 804, 983

Farancia abacura (mud snake)
Although common in the Savannah River swamp system and various aquatic habitats on the SRS, a few individuals of this species have been collected in Par Pond. The majority of captures on the SRS have been in the spring or early summer months. March through May is the period of greatest activity.

At least some females apparently lay eggs a long distance from water in the sandy upland areas, and the young, after hatching (whether in fall or spring), do not proceed to the aquatic habitat until winter has passed. Juveniles at Ellenton Bay have been observed eating or disgorging larval *Ambystoma talpoideum*. Although this genus is reported to eat American eels and *Amphiuma*, neither of these prey items is known to occur in the

Ellenton Bay habitat, despite more than two decades of intensive sampling for aquatic organisms.

Smaller specimens of *F. abacura* can be collected in drift fences or minnow traps. Large specimens are sometimes captured in shallow aquatic areas at night. SRS references 396, 566, 643, 804, 1210

Farancia erytrogramma (rainbow snake)
Farancia erytrogramma is found in or around various aquatic habitats. Most specimens have come from minnow traps at Steel Creek or from Risher Pond and Ellenton Bay drift fences and minnow traps. The majority of captures on the SRS have been in the spring, especially March or April, or early summer months. None has been collected in November or December. Ellenton Bay supports a large population of *F. erytrogramma* but, similar to *F. abacura*, they have been known to eat only *A. talpoideum* larvae. Mean SVL of males is 62 cm, and of females, 92 cm. Smaller specimens can be collected in drift fences or minnow traps. Large specimens are sometimes captured in shallow aquatic areas at night. SRS references 396, 543, 566, 643, 728

Heterodon platyrhinos (eastern hognose snake)
This species is found in sandy turkey oak and longleaf pine sand hill regions and abandoned old fields anywhere these habitats occur on the SRS. Individuals are frequently found on roads, but during daytime only. This species has been collected every month of the year at the SRS, but most commonly in April through October.

The diet is predominantly toads (*Bufo terrestris*), but they have been observed to eat *Scaphiopus*, *Rana*, and *Notophthalmus*. Mean SVL and body mass of males are 49 cm and 103 g, and 58 cm and 170 g for females. Numerous specimens are captured on highways and in drift fences.

A large proportion of *H. platyrhinos* are melanistic on the SRS, whereas the *H. simus* color pattern varies little. At Ellenton Bay, where a large number of *H. platyrhinos* have been collected, all of the adults have been melanistic. Melanism develops sometime after the first year of growth as all juveniles have had well-developed color patterns. SRS references 396, 439, 475, 566, 624, 643, 804, 1133

Heterodon simus (southern hognose snake)
This species is restricted to sandy turkey oak and longleaf pine sand hill regions. They are found in habitats similar to those in which the eastern hognose occurs but are less common. Specimens have been collected from December through March, but most are collected between April and October. The diet is predominately toads (*Bufo terrestris*). Mean SVL and

body mass of males are 33 cm and 46 g, and of females, 47 cm and 120 g. Numerous specimens are captured on highways and in drift fences. This species is never melanistic on the SRS. SRS references 396, 566, 643, 804, 1133

Lampropeltis getulus (common kingsnake)
This kingsnake is common, particularly in low-lying habitats around aquatic areas. Several individuals are encountered each year around the Ellenton Bay drift fence. It occurs in a wide variety of habitats on the SRS, often in the vicinity of permanent or temporary aquatic areas. Although concentrations are known to occur in selected areas such as Ellenton Bay, normally no more than one or two individuals are encountered during any one day. The species is most active between April and June and is seldom collected from December to March.

On at least two occasions individuals have been observed probing into turtle nests, presumably to eat the eggs. One captured individual regurgitated turtle eggs. Mean SVL and body mass of males are 121 cm and 659 g, and of females, 111 cm and 365 g. Common kingsnakes are occasionally captured at drift fences or sometimes by hand. SRS references 152, 396, 397, 439, 555, 566, 643, 804, 1073, 1133

Lampropeltis triangulum (scarlet kingsnake)
The scarlet kingsnake is restricted primarily to areas of sandy soil and open pine woods on the SRS. Several individuals were collected on Road F at night during late summer 1977. Three specimens were caught in drift fences around Pond C, and occasional specimens have been captured at Rainbow Bay, Sun Bay, Ellenton Bay, and Dry Bay. Those captured on roads have been found at night. Several have been found beneath pine bark and in pine stumps from March through October. Little is known about the ecology of this species on the SRS. Two females are known to have laid three eggs each in May (Tony Mills, pers. comm.). SRS references 566, 643

Masticophis flagellum (coachwhip)
Coachwhips are primarily associated with dry, sandy habitats such as abandoned old fields and scrub oak forest areas but have been reported from throughout the SRS; they may occur in pine or hardwood areas on occasion and are strictly diurnal. Active individuals are seen primarily from May through September. Their diet consists predominantly of lizards, small mammals, and birds. Mean SVL and body mass of males are 138 cm and 501 g, and of females, 123 cm and 344 g. Most specimens are captured by hand, although small specimens have been taken at drift fences in pitfall or funnel traps. SRS references 396, 397, 439, 566, 643

Nerodia cyclopion (green water snake)
Large populations occur in Ellenton Bay and Dry Bay, and several specimens have been collected in Par Pond and Pond B. None has been collected in stream or swamp habitats on the SRS. Green water snakes have been captured in every month except January and October. Mean SVL and body mass of males are 59 cm and 154 g, and 83 cm and 438 g for females. Five pairs were found in copulation at Ellenton Bay during warm days in February 1977 and 1978. Minnow traps, hand collecting, and drift fences have all been effective capture techniques. SRS references 209, 566, 643

Nerodia erythrogaster (red-bellied water snake)
Nerodia erythrogaster is found terrestrially more often than the other *Nerodia* species and is encountered in and around both large and small aquatic habitats, including streams, throughout the SRS. Individuals may be encountered several hundred meters away from permanent water. The species is not abundant at any single location. Most individuals have been captured between March and July. Mean SVL and body mass of males are 77 cm and 286 g, and of females, 99 cm and 516 g. Hand capturing has been the most effective technique. SRS references 396, 439, 566, 643, 804

Nerodia fasciata (banded water snake)
The banded water snake is at least present and often common in almost all permanent or temporary aquatic habitats on the SRS. Individuals have been captured from March to September, with occasional records for every month of the year. Mean SVL and body mass of males are 54 cm and 114 g, and of females, 64 cm and 247 g. This species gives birth in late summer. Minnow traps, hand collecting, and drift fences have all been effective techniques for capture.
The northern water snake (*N. sipedon*) may be present on the site also, but detailed taxonomic studies have not been conducted. SRS references 566, 624, 643, 804, 815

Nerodia taxispilota (brown water snake)
The brown water snake is associated with most stream, river, and swamp habitats on the SRS but is seldom found in Carolina bays. It is occasionally encountered in Par Pond. Large numbers can be found along Upper Three Runs Creek and in the Steel Creek delta. Mean SVL and body mass for males are 65 cm and 210 g, and for females, 83 cm and 718 g. This species gives birth in late summer. The diet is believed to be primarily fish. The largest numbers are captured by hand during the daytime in spring and early summer in stream and swamp habitats. In contrast to other *Nerodia*

on the SRS and elsewhere, this species does not appear to be active nocturnally. SRS references 276, 326, 396, 397, 439, 566, 589, 624, 643, 815, 983

Opheodrys aestivus (rough green snake)
This species is associated with thickly vegetated areas having vines, bushes, and shrubs, throughout the SRS. They have been collected from May to September. Although only a few have been collected on roads and at drift fences, 155 individuals were caught during the flooding of Steel Creek bottomlands to create L-Lake (October 1985). Except for the unusual sample from Steel Creek, most specimens have been captured singly either crossing roads or arboreally. *Opheodrys* is almost exclusively insectivorous. SRS references 396, 566, 643

Pituophis melanoleucus (pine snake)
This species is restricted to sandy habitats, especially old fields, and turkey oak–pine forests, where it is fossorial much of the time. Most species are found from June to October. Gravid females have been collected in late spring, and the young appear in early autumn. Mean SVL and body mass of males are 107 cm and 460 g, and of females, 106 cm and 443 g. Most specimens are obtained incidentally as captures on highways during daylight hours. SRS references 396, 439, 566, 643, 804

Regina rigida (glossy crayfish snake)
Two specimens have been collected from the drift fence at Risher Pond, and a DOR (dead on road) specimen was found on S.C. 125 near Steel Creek in 1984. In March 1988 one specimen was collected in a minnow trap at Steel Creek Bay and another at Bay 23 (across highway and about 2 km north of Dry Bay). Additional captures were made in 1988 and 1989 at Squirrel Bay by David Scott. The diet on the SRS is unknown, although crawfish, reportedly the primary food, have been present at all sites. Drift fences or minnow traps are presumably the most effective collecting techniques. SRS reference 643

Regina septemvittata (queen snake)
The first specimen on the SRS was captured in April 1983 by Chuck Vincent in one of the small tributaries to Upper Three Runs Creek on the northeastern edge of the SRS. No other specimens were recorded until October 1986 when John Aho collected five specimens from minnow traps placed in the same creek. SRS references, none

Rhadinaea flavilata (yellow-lipped snake)
The first specimen on the SRS was collected by hand near Dry Bay on 9 May 1987. Later in 1987, another specimen was captured in a pitfall trap at Dry Bay by Tony Mills, and David Scott captured two individuals in pitfall traps at Squirrel Bay. In 1988 Tony Mills collected another by hand at Squirrel Bay. Previous published records for this species in South Carolina are confined to the lower coastal plain (Myers, 1967), but Rudy Mancke is reported to have collected a specimen in lower Richland County in 1980. A total of eight specimens have been captured on the SRS. SRS references 566, 1227

Seminatrix pygaea (black swamp snake)
The concentration of black swamp snakes at Ellenton Bay is the largest known in South Carolina, and nearly all of the available information for this species on the SRS comes from that population. Specimens have been collected also at two other standing-water habitats on the SRS: Dry Bay and below the Risher Pond dam.

Black swamp snakes have been collected at Ellenton Bay in every month except January and February. Water temperature seems to be important in determining diel activity patterns. *Seminatrix* were rarely trapped if the minimum water temperature was below 16°C. Seasonal activity patterns are similar to those of many snakes — males first appear in late March or early April; the peak period of female activity is a month or so later, in May or June.

Activity is concentrated along the shoreline and edges of emergent vegetation. In over 300 trap-nights in open water only a single specimen was captured, compared with dozens of captures along the shoreline. The 225 captured showed a sex ratio of 1:1 (118 males:107 females). Females produce their first clutch when 24–26 cm SVL and give birth to 5–15 young in August or September.

At Ellenton Bay, black swamp snakes are primarily piscivorous. *Gambusia affinis* made up 82% of 124 prey items removed from specimens' guts. Leeches (10%), *Ambystoma talpoideum* (9%), and earthworms (2%) were also ingested. Prey were eaten headfirst generally, and multiple prey were sometimes recovered from single individuals.

Females were longer (mean SVL = 32.2 cm) and heavier (mean mass = 19.7 g) than males (mean SVL = 29.2 cm, mean mass = 15.3 g).

Minnow traps and drift fences have been highly effective at Ellenton Bay. Specimens may sometimes be found under debris (e.g., tin, boards, etc.) or in vegetation at the water's edge. A single small specimen was collected by ripping up sod with a potato rake. This technique might be productive, especially if the bay dries and snakes concentrate in the remaining wet areas, but it is extremely labor intensive. (This account was contributed by Ray Loraine.) SRS references 566, 643, 804

Storeria dekayi (brown snake)
Occasional individuals are found in a diversity of habitats from moist to dry woodlands to swampy areas if abundant ground cover and litter are available. Brown snakes have been collected every month of the year at SRS, but May through October seems to be the most active period. Females collected on the SRS have a mean SVL of 15 cm. Drift fences with pitfall traps have been the most effective collecting technique. SRS references 566, 643, 804, 1133

Storeria occipitomaculata (red-bellied snake)
This is one of the few species active during all months of the year, with the peak of activity during September and October. Specimens are occasionally collected on SRS roads at night. Drift fences in moist woodland areas with abundant ground litter around Pond C, Lost Lake, and the old SREL site revealed the presence of this species. Occasional individuals have been found in a wide variety of SRS habitats, usually wet or damp, though not necessarily near permanent aquatic areas. Mean SVL and body mass of both males and females are 15 cm and 2 g. Red-bellied snakes feed primarily on slugs. Drift fences with pitfall traps are the most effective collecting technique. SRS references 566, 643, 804, 885, 1133

Tantilla coronata (southeastern crowned snake)
This snake is found in a variety of habitats but mostly in wooded areas with abundant ground litter and debris. They have been collected most commonly from April to September. Mean SVL and body mass of both sexes are 19 cm and approximately 3 g. Most SRS captures have been at drift fences in upland wooded sites.

This species is the most frequently captured snake in pitfall traps on the SRS. More than 600 specimens have been collected from a variety of study areas, primarily associated with open or sparsely wooded habitats. Crowned snakes on the SRS feed primarily on centipedes. SRS references 566, 643, 728, 804, 1133

Thamnophis sauritus (eastern ribbon snake)
The few individuals reported have been found in association with aquatic areas in various parts of the SRS. They have been collected in every month but January and August. Drift fences with pitfall traps have been the most effective collecting technique. SRS references 396, 566, 643, 728, 804

Thamnophis sirtalis (common garter snake)
Occasional individuals have been found in several different SRS habitats, usually wet or damp, though not necessarily near permanently aquatic areas. They have been collected every month but January, and most were collected from July to October. Although the species has been encountered sporadically throughout the SRS, no large concentrations have been discovered. Drift fences with pitfall traps have been the most effective collecting technique. SRS references 392, 396, 566, 643, 804, 812, 903, 1133

Virginia striatula (rough earth snake)
The rough earth snake went undiscovered for many years, and it has only been found at three localities on the SRS. Hand collecting at Gun Site 51 west of Road A yielded a dozen specimens in the spring of 1982 — the largest sample of this species taken in the region. Whether this species occurs in isolated populations that have for the most part gone undetected or whether it is a widely distributed form that can only be collected in large numbers at certain times of the year is unknown at this time. Its abundance in many areas of the country leads us to believe that it is truly an uncommon species on the SRS. They have been collected only in May, July, and September. Several have also been collected at Dry Bay (approximately 3 km away from Gun Site 51) and at Rainbow Bay, mostly caught by hand beneath debris or in pitfall traps. SRS references 566, 804

Virginia valeriae (smooth earth snake)
Most individuals have been collected with drift fences, the largest numbers from the east side of Pond C and from Lost Lake. May–September is the period of greatest observed activity. They have not been collected in January, February, or December. Mean SVL of males and females is about 13 cm. Drift fences have been the most effective capture technique. SRS references 566, 643, 804, 1133

Venomous Snakes

Agkistrodon contortrix (copperhead)
The copperhead is more nearly ubiquitous, but is a less common snake on the SRS, than the cottonmouth. It may be encountered in any terrestrial habitat. Most specimens have been collected on roads from April to September in several different habitat types, ranging from strictly terrestrial uplands to the margins of lowland aquatic areas, including the Savannah River swamp. Mean SVL and body mass of males are 76 cm and 288 g, and of females, 67 cm and 220 g. Hand capture of individuals on highways or in terrestrial habitats is the best collecting technique. Although winter

concentrations of this species are frequently encountered in other parts of its range, none has been discovered on the SRS. SRS references 396, 439, 566, 643

Agkistrodon piscivorus (cottonmouth)
The cottonmouth, or water moccasin, is one of the most frequently encountered and abundant species of snakes in aquatic habitats on the SRS. Cottonmouths are found in association with almost every wet habitat on the SRS, with the notable exception of Par Pond. Concentrations of a dozen or more are occasionally encountered in the Savannah River swamp area and in some Carolina bays (e.g., Flamingo Bay) during summer when water levels are low. They are common in some areas from March to November and are occasionally collected during December, January, and February.

Mean SVL and body mass of males are 89 cm and 710 g, and of females, 73 cm and 466 g. Several cottonmouths were found gorging themselves on bullfrog tadpoles and *A. talpoideum* larvae at Flamingo Bay as it was drying in 1981. Individuals are readily captured by hand around many aquatic habitats on the SRS.

The Ellenton Bay records are of particular interest because they are so recent. Although the area has been studied intensively both with drift fences and frequent daytime and nighttime collecting efforts since 1967, none was seen there until 1981 when three juveniles were collected entering the aquatic portion of the bay (based on their collection in pitfall traps on the outside of the fence). Whether these individuals and additional entries in later years will result in the establishment of a population at this site remains to be seen, but the potential for studying colonization of a habitat by a snake species that was formerly absent is noteworthy. The rarity of cottonmouths in the Par Pond reservoir system is also notable. In more than 20 years of intensive study, very few cottonmouths have been seen in any of the three large reservoirs. Since the margins of Par Pond are contiguous with numerous habitats suitable for cottonmouths, we attribute their scarcity to large populations of alligators (T. M., Murphy, 1977) and largemouth bass (Gibbons et al., 1978) inhabiting the lake. We assume that one aspect of this conspicuous snake's behavior (slow surface swimming) is incompatible with maintenance of appreciable numbers in an environment with such high predatory pressure. SRS references 305, 396, 566, 643, 804

Crotalus horridus (timber rattlesnake; canebrake rattlesnake)
This species is found in a wide variety of terrestrial habitats on the SRS. Few *Crotalus* are seen before mid-May in most years, but active individuals

can be encountered as late as November. Mean SVL and body mass of males are 111 cm and 1091 g, and of females, 110 cm and 929 g. Most individuals have been collected on roads, primarily in early evening or at night.

The canebrake rattlesnake is the only large rattlesnake on the SRS. The eastern diamondback rattlesnake (*C. adamanteus*) is not found this far inland and has not been reported within 80 km of the southern border of the SRS. SRS references 301, 305, 396, 439, 566, 643, 804

Micrurus fulvius (eastern coral snake)

Fewer than a dozen coral snakes have been found on the SRS, all within the northwestern sector and mostly in autumn (September and October). All have been associated with turkey oak–pine habitats. A female collected near Aiken laid four elongate eggs in captivity in June 1982 (Laurie Vitt, pers. comm.). All individuals have been collected on roads, except for a single capture at a pitfall trap at Lost Lake. SRS references 566, 643

Sistrurus miliarius (pygmy rattlesnake)

Most SRS specimens have been encountered in turkey oak–longleaf pine forest habitats in the northern sector on roads at night during late summer, primarily in August and September. Several individuals were captured by road collecting at night during late summer 1977, and a few have been captured at drift fences with pitfall traps at the SREL Food Field. SRS references 566, 643

Problem Species

Investigators might encounter special problems when studying SRS reptiles and amphibians. These include venomous species, unresolved records and unverified reports or sightings, possibilities of encountering endangered or introduced species, and the possibility that edible species could have individuals contaminated by local pollutants, including radionuclides.

Unresolved Records

Reports of some species of reptiles and amphibians from the SRS over the past 30 years have not been confirmed by subsequent investigators (table 2). Possible reasons include misidentifications or cataloguing errors, the existence of previously existing populations that are no longer present, inadequate sampling in particular habitats or sites in recent times, and problems associated with taxonomic assignment and interpretation. The following SRS species or species combinations are presently in question:

Desmognathus fuscus (dusky salamander)
Adults of this species are difficult to distinguish from *D. auriculatus* in many instances. Few *Desmognathus* from the site have been closely examined, and both species may be present.

Bufo woodhousei (Woodhouse's toad; Fowler's toad)
Five individuals were reported by Freeman (1956). Julian Harrison (pers. comm.), upon reexamination of a preserved specimen in the Charleston Museum, has confirmed the species as *B. woodhousei*. This species has not been reported from the SRS since the early records of the 1950s. Verification of its current existence on the SRS is needed.

Table 2. Total numbers of amphibian and reptile species on the SRS (including species that have been reported but are of questionable occurrence).

	Families	Genera	Species
Amphibians			
Salamanders	6	9	17
Frogs and Toads	5	7	26
Total	11	16	43
Reptiles			
Crocodilians	1	1	1
Turtles	4	10	12
Lizards	4	6	9
Snakes	3	23	36
Total	12	40	58
Total	23	56	101

Hyla versicolor (gray treefrog)
This species may be present in several areas on the SRS but because of its phenotypic similarity to *H. chrysoscelis* may not have been recognized as *H. versicolor*. An examination of karyotypes from several populations might provide definite evidence that both species occur commonly on the site.

Pseudacris brimleyi (Brimley's chorus frog)
Humphries (1953–54) reported this species from the northern perimeter of the SRS but no voucher specimens exist. No individuals have been reported since the 1950s, but an intensive search in the northernmost sections of the site might reveal small populations.

Pseudacris triseriata (striped chorus frog)
The presence of this species on the SRS is problematical. Because of its similarity to *P. nigrita* in call and appearance, its status on the site is uncertain at this time.

Nerodia sipedon (northern water snake)
The banded water snake (*N. fasciata*) definitely occurs on the SRS and represents the more commonly encountered species of the two. However, occasional individuals are collected that could be assigned morphologically to *N. sipedon*. The verification of its occurrence will require a detailed examination of a series of specimens.

Several reptile and amphibian species of South Carolina or Georgia have geographic ranges that encompass or closely approach the SRS, but the species have not been reported from the site. The presence or absence of some species is problematical due to similarity of appearance with another species or lack of a recent verification. These problem species include:

AMPHIBIANS

Salamanders

Ambystoma cingulatum (flatwoods salamander)
Ambystoma mabeei (Mabee's salamander)
Desmognathus fuscus (dusky salamander)
Hemidactylium scutatum (four-toed salamander)
Plethodon websteri (Webster's salamander)
Pseudobranchus striatus (dwarf siren)
Stereochilus marginatus (many-lined salamander)

Frogs and Toads

Bufo americanus (American toad)
Bufo woodhousei (Woodhouse's toad)
Hyla andersoni (pine barrens treefrog)
Hyla versicolor (gray treefrog)
Pseudacris brimleyi (Brimley's chorus frog)
Pseudacris triseriata (striped chorus frog)
Rana heckscheri (river frog)

REPTILES

Snakes

Crotalus adamanteus (eastern diamondback rattlesnake)
Lampropeltis calligaster (mole snake)
Nerodia sipedon (northern water snake)

Lizards

Eumeces egregius (northern mole skink)

Turtles

Gopherus polyphemus (gopher tortoise)
Trionyx ferox (Florida softshell)

Fall Line Subspeciation

The northern border of the SRS is only 40 km south of the Fall Line, which parallels the presumed zone of intergradation of several subspecies of reptiles and amphibians. Specific research efforts are needed to establish the subspecific designations of many of the reptiles and amphibians on the SRS. Collections of SRS specimens have been provided to the following museums and could be used for this purpose: University of Georgia, Clemson University, George Mason University, Carnegie Museum, College of Charleston and Charleston Museum, Smithsonian Institution, and South Carolina State Museum. In addition, a largely regional herpetological collection with over 3000 specimens is maintained by SREL.

Venomous Species

Five species of poisonous snakes have been reported from the SRS. The diamondback rattlesnake (*Crotalus adamanteus*) does not occur on the site, and the pygmy rattlesnake (*Sistrurus miliarius*) and eastern coral snake (*Micrurus fulvius*) are seldom encountered. The other three species of southeastern poisonous snakes are more common. The cottonmouth (*Agkistrodon piscivorus*) is most frequently encountered in aquatic situations, whereas the canebrake rattlesnake (*Crotalus horridus*) and the copperhead (*Agkistrodon contortrix*) are the most common on land. The potential for serious snakebite exists on the site. Because of the proximity of the SRS to medical facilities in Augusta, Georgia, or Aiken, South Carolina, professional medical attention is ordinarily warranted in lieu of extensive first-aid measures.

Edible Species

The combination of a species being edible by humans and the potential for uptake of contaminants by the species could result in unfortunate consequences. It is not the intent of this report to indicate where on the SRS such situations might arise, only to indicate the species in the region that are generally considered edible. These include the bullfrog (*R. catesbeiana*), pig frog (*R. grylio*), and five turtles (*C. serpentina*, *T. scripta*, *P. floridana*, *T. spiniferus*, and *D. reticularia*). Although many of the snakes are large enough for human consumption, as well as palatable, no species in this region is habitually eaten, although *C. horridus* has been used for food in some parts of the country.

Introduced Species

The practice of keeping and eventually releasing exotic pets has led to the introduction of new species in many areas. The horned toad (*Phrynosoma cornutum*) is an established species in some sections of the South Carolina coastal plain, such as Isle of Palms and Sullivan's Island (Julian Harrison, pers. comm.). It has not been found on the SRS, but the possibility exists that it could become established because of the abundant sandy habitats. Also, the Mediterranean gecko (*Hemidactylus turcicus*) has been found in Augusta, Georgia (approximately 30 km from the SRS), by Tony Mills and could conceivably be present in some areas on the SRS.

Endangered Species

Special state and federal regulations protect some species with declining or small populations. These animals cannot be intentionally killed and can be collected only by permit from the state and federal government. Only one federally protected reptile species, *Alligator mississippiensis*, is known to occur on the SRS, being found in the Par Pond reservoir system, many lakes, and the Savannah River swamp.

Reptiles or amphibians should not be collected for research purposes on the SRS without a permit issued by the South Carolina Wildlife and Marine Resources Department, Columbia, S.C., and authorization by the Savannah River Operations Office of the U. S. Department of Energy.

Herpetological Research Problems

The list of herpetological research questions on SRS species is virtually endless. Certain problems are identifiable as a consequence of previous research on the site and deserve special mention as potential study projects. For example, *Kinosternon bauri* is known to occur on the SRS at certain locations, but its distribution on the site, or even in South Carolina, has not been established. The association and interaction of *K. bauri* with *K. subrubrum* has not been determined at all. Situations like this in which certain species are easily confused with others is a common problem. Determining the presence and the association or interaction of similar-appearing species would be worthwhile from the standpoint of firmly establishing which species of reptiles and amphibians definitely occur on the site. Examples are: *Desmognathus fuscus, Pseudacris brimleyi, Bufo woodhousei, Rana heckscheri, Hyla versicolor, Pseudacris triseriata,* and

Nerodia sipedon, all of which have close relatives on the site with which they might be easily confused by nonherpetologists and experts alike.

In addition to problems associated with potentially confusable species, several species have ranges that approach or encompass the SRS, but their presence has not been confirmed. For example, *Plethodon websteri* occurs in the Stevens Creek area above Aiken, S.C., and the range of *Rana heckscheri* approaches the southern border of the SRS. *Lampropeltis calligaster*, whose geographic range supposedly encompasses the SRS, has been documented from the site only by two questionable specimens. One was an individual from Ellenton Bay that escaped before confirmed identification; the other is a specimen captured by Dr. Ronn Altig on 21 July 1969 and now deposited in the University of Michigan Museum of Zoology. Some question exists of whether the individual was an escapee from a caged collection being kept near the capture site at SREL. This species is fossorial and nocturnal (Mount, 1975; Conant, 1975), and the frequent occurrence of individuals within 35 km of the northern boundary has led to speculation that the species will eventually be discovered on the SRS. However, although the species is secretive and generally collected only during early spring, the use of drift fences around old-field habitats and extensive road collecting presumably would have revealed its presence if populations were present on the SRS. On the other hand, the recent discovery of *R. flavilata* and the reconfirmations of *A. maculatum*, *R. rigida*, *R. grylio*, and *R. palustris* after more than 10 years without sightings proclaims the possibility that a population may eventually be found.

The numbers of species of reptiles and amphibians reported on the SRS have varied since the initial herpetological surveys of Freeman. The numbers in the following list indicate how many species were reported in different publications as confirmed inhabitants of the SRS. The species names in parentheses designate those that were added from the most recently published report. Amphibian lists were not reported until 1978. The number of frog species listed by Freeman (1955) includes only those species confirmed in later surveys.

Freeman, 1955
9 turtles; 8 lizards; 27 snakes; 22 frogs; 14 salamanders

M. J. Duever, 1967 (thesis)
10 turtles (*K. bauri*); 9 lizards (*O. attenuatus*); 30 snakes (*T. sirtalis, L. triangulum, V. valeriae*)

Gibbons, 1977
10 turtles; 9 lizards; 31 snakes (*R. rigida*)

Gibbons and Patterson, 1978
    11 turtles (*P. concinna*); 9 lizards; 31 snakes; 26 frogs
    (*P. triseriata, P. ocularis, R. palustris, R. areolata*); 16 salamanders
    (*S. lacertina, A. means*)

Gibbons and Semlitsch, 1991
    12 turtles (*C. picta*); 9 lizards; 36 snakes (*V. striatula, C. amoenus, R. flavilata, R. septemvittata*); 26 frogs; 17 salamanders
    (*A. maculatum*)

A wealth of ecological problems abounds on the SRS. Mentioning each would be too lengthy a process, but many herpetological research questions can be posed on the basis of the numerous publications on the reptiles and amphibians of the SRS.

Bibliography

Many SRS reptiles and amphibians have been extensively studied, and many publications have resulted. The following bibliography provides categorical references for herpetological studies conducted on the SRS.

Papers included in the SREL Herpetological Reprints were based on reptiles and amphibians of the SRS. Each paper has been assigned to subject categories as listed and explained below. We have developed a computer program to search the data base and print cross-indexed bibliographies. The search program and/or the complete data base can be made available to investigators currently working on these species. The articles that apply to particular categories are listed at the back of the bibliography. These listings also give all of the SRS information that applies to a particular species. SREL numbers are those on the laboratory's distribution list. Available reprints may be obtained upon request.

We have attempted to compile a complete bibliography for the SRS species. Users of this bibliography are encouraged to supply us with any missing references and to comment on the format so that the bibliography can be updated and revised to make it more useful to researchers. We hope to keep the bibliography updated and produce computer printouts upon request.

Subject Categories for SREL Herpetological Reprints

Numbers following each category represent SREL reprint numbers (see page 110).

1 Habitat: Papers describing the habitat and associated species for the reptiles and amphibians of the SRS. These include the effects of habitat on the biology of particular species.

0048	0199	0200	0209	0216	0244	0279	0305	0316
0318	0326	0362	0381	0384	0391	0392	0396	0405
0429	0520	0543	0565	0566	0570	0589	0590	0612
0624	0629	0630	0643	0658	0664	0666	0675	0676
0677	0678	0684	0706	0710	0716	0717	0728	0775
0779	0793	0836	0840	0848	0858	0863	0866	0868
0869	0885	0963	0972	1007	1044	1091	1118	1120
1188	1205	1210	1244	1301				

2 Distribution: Papers that give specific geographic locations that can be used to describe the range of a species.

0048	0079	0152	0199	0200	0209	0216	0222	0227
0244	0262	0272	0279	0301	0305	0315	0316	0318
0326	0362	0381	0384	0391	0392	0396	0397	0405
0429	0439	0475	0520	0543	0553	0565	0566	0570
0589	0590	0612	0624	0629	0630	0643	0658	0662
0664	0666	0675	0676	0677	0678	0682	0684	0701
0706	0716	0718	0728	0743	0744	0775	0779	0793
0804	0816	0836	0840	0858	0863	0866	0868	0869
0875	0882	0885	0886	0903	0908	0963	1091	1227
1244	1301							

3 Growth and Development: Papers on size- or age-specific changes in dimensions or other characteristics.

0048	0199	0200	0216	0305	0384	0392	0475	0507
0520	0565	0570	0590	0612	0662	0664	0666	0677
0684	0718	0792	0840	0848	0852	0868	0885	0972
1050	1120	1205	1210	1222	1244	1301		

4 Body Composition: Papers that deal with attributes of body composition such as caloric value, body fat, and protein.

0199	0262	0276	0280	0301	0302	0305	0397	0475
0507	0589	0662	0827	0852	0999			

5 Reproduction: Papers concerning reproductive characteristics such as percentage of pregnant females, litter and clutch size, embryo mortality, parental investment, and hatchling emergence.

0048	0200	0209	0216	0280	0305	0475	0520	0543
0565	0570	0589	0590	0603	0612	0652	0662	0664
0666	0675	0676	0677	0718	0728	0779	0782	0812

0815	0816	0827	0848	0852	0858	0863	0866	0868
0869	0885	0908	0963	0993	1008	1022	1118	1120
1151	1205							

6 Mortality and Age Structure: Papers dealing with sources of mortality, amounts of mortality in relation to various factors, and resulting age structure and sex ratio.

0048	0200	0216	0305	0392	0507	0520	0543	0666
0675	0676	0677	0684	0716	0728	0775	0793	0840
0848	0863	0868	0885	0963	1050	1118	1127	1222

7 Numbers and Density: Papers dealing with relative abundance or numbers of reptiles and amphibians per unit area.

0048	0079	0209	0216	0316	0362	0391	0590	0643
0658	0666	0675	0684	0728	0804	0863	0866	0868
1007	1118	1205	1210	1280	1301			

8 Genetics: Papers concerning determination of marker genes, gene frequencies and inheritance, or genetic similarity of various forms indicated by electrophoretic or experimental breeding studies.

0664   0701   1044   1260

9 Movement Patterns: Papers concerning dispersal, migration, or movement within a home range.

0048	0200	0209	0244	0272	0279	0316	0318	0362
0391	0405	0520	0543	0565	0570	0590	0658	0664
0675	0677	0684	0716	0744	0804	0863	0866	0868
0885	0908	0964	1019	1051	1118	1133	1301	

10 Behavior: Papers concerning all behavior other than movement.

0048	0152	0200	0209	0244	0272	0279	0315	0316
0318	0555	0565	0710	0716	0728	0742	0775	0804
0836	0844	0866	0869	0886	0903	1008	1011	1031
1032	1033	1050	1051	1056	1063	1064	1073	1095
1102	1110	1118	1120	1139	1143	1160	1162	1164
1169	1184	1205	1210	1211	1244	1280	1282	1297
1309	1326	1336						

11 Physiology: Papers dealing with all physiological attributes with the exception of those concerning metabolism or energetics.

0262	0280	0475	0507	0624	0662	0678	0706	0710
0742	0744	0836	0885	0886	0903	0963	0972	0993
0999	1033	1056	1064	1078	1095	1102	1111	1112
1143	1169	1184						

12 Metabolism and Energetics: Papers dealing with oxygen consumption, carbon dioxide production, food consumption or caloric requirements.

0199	0280	0384	0475	0624	0664	0742	0812	1007
1050	1063	1073	1268	1342				

13 Parasites and Diseases: Papers that identify or describe parasites or diseases.

0227   0381   0405   0629   0630

14 External Morphology: Papers that include measurements of external dimensions or descriptions of external appearance not specifically related to changes with age.

0222	0439	0543	0590	0718	0728	0844	0875	0882
0963	1011	1022						

15 Internal Morphology: Papers dealing with attributes of internal morphology such as internal anatomy, cytology, histology, and karyology

0222   0852   0882   1151

16 Taxonomy and Systematics: Papers dealing with nomenclature, placement of populations in higher taxonomic categories, and subspecific designations within the species.

0222   0875   0882

17 Paleontology: Papers that describe the fossil occurrence of SRS reptiles or amphibians or evaluate methods for identifying fossils of these species.

0658   0664   0722   0743   0804

18 Radioecology: Papers that describe effects of radiation and the use of radioisotopes.

0244	0318	0396	0397	0429	0475	0603	0652	0716
0743	0972	0983	1112	1260	1268	1304		

19 Evolution: Papers that discuss the evolutionary biology of particular species including modes of selection.

0565 0664 0815 1095 1151 1342

20 Wildlife Management: Papers that present data of management interest or that present different management plans for populations on the SRS.

0553	0658	0664	0710	0722	0743	0804	1078	1091
1307								

21 Bibliography: Papers that contain extensive literature-cited sections or that are primarily bibliographical compilations.

0222 0812 0821

22 Collection and Research Techniques: Papers that describe particular sampling or collecting techniques.

0048	0199	0200	0209	0216	0244	0262	0263	0272
0276	0279	0301	0302	0305	0315	0316	0318	0362
0384	0391	0392	0396	0397	0429	0439	0440	0475
0520	0543	0553	0555	0565	0570	0589	0590	0603
0612	0624	0643	0658	0662	0666	0675	0677	0678
0684	0701	0706	0716	0722	0728	0742	0743	0744
0775	0779	0792	0793	0804	0815	0816	0827	0836
0840	0844	0852	0858	0863	0866	0868	0882	0885
0886	0892	0903	0963					

Species of Reptiles and Amphibians for Bibliography

Papers that contain significant information about particular species of the SRS, as indicated by numbers after the species accounts.

ORDER CAUDATA (Salamanders)
1 *Necturus punctatus* (dwarf waterdog): 0643.
2 *Amphiuma means* (two-toed amphiuma): 0643, 0840.

3 *Siren intermedia* (lesser siren): 0643, 0840.
4 *Siren lacertina* (greater siren): 0643.
5 *Ambystoma maculatum* (spotted salamander): 0643, 0716, 0844, 0863, 1162, 1244.
6 *Ambystoma opacum* (marbled salamander): 0590, 0643, 0658, 0804, 0844, 1280.
7 *Ambystoma talpoideum* (mole salamander): 0276, 0590, 0643, 0658, 0675, 0684, 0716, 0743, 0804, 0844, 0972, 1120, 1162, 1164, 1205, 1244, 1280, 1301.
8 *Ambystoma tigrinum* (tiger salamander): 0276, 0590, 0643, 0716, 0804, 0844, 0863, 0964.
9 *Notophthalmus viridescens* (eastern or red-spotted newt): 0643, 0658, 0804, 0848, 1280.
10 *Desmognathus auriculatus* (southern dusky salamander): 0643, 0840, 0848.
11 *Eurycea cirrigera* (two-lined salamander): 0643, 0677, 0743, 0804, 0840, 0848.
12 *Eurycea longicauda guttolineata* (long-tailed or 3-lined salamander): 0276, 0643, 0840, 0848.
13 *Eurycea quadridigitata* (dwarf salamander): 0643, 0658, 0675, 0677, 0706, 0804, 1280.
14 *Plethodon glutinosus* (slimy salamander): 0643, 0658, 0666, 0804, 0848.
15 *Pseudotriton montanus* (mud salamander): 0643, 0804, 0840.
16 *Pseudotriton ruber* (red salamander): 0643, 0804, 0840.

ORDER ANURA (Frogs and toads)
17 *Scaphiopus holbrooki* (eastern spadefoot toad): 0222, 0276, 0280, 0362, 0643, 0658, 0792, 0804.
18 *Bufo quercicus* (oak toad): 0222, 0279, 0280, 0362, 0643, 0658, 0804.
19 *Bufo terrestris* (southern toad): 0222, 0276, 0280, 0362, 0392, 0475, 0643, 0658, 0664, 0804.
20 *Acris crepitans* (northern cricket frog): 0222, 0643.
21 *Acris gryllus* (southern cricket frog): 0222, 0276, 0279, 0280, 0362, 0643, 0804.
22 *Hyla avivoca* (bird-voiced treefrog): 0222, 0643.
23 *Hyla chrysoscelis* (Cope's gray treefrog): 0222, 0643, 0804, 1222.
24 *Hyla versicolor* (gray treefrog): 0222, 0276, 0643, 0658.
25 *Hyla cinerea* (green treefrog): 0222, 0276, 0362, 0429, 0643, 0658, 0804.
26 *Hyla femoralis* (pine woods treefrog): 0222, 0262, 0276, 0362, 0643, 0804, 1301.
27 *Hyla gratiosa* (barking treefrog): 0222, 0280, 0643, 0804.
28 *Hyla squirella* (squirrel treefrog): 0222, 0276, 0280, 0362, 0643, 0658, 0684, 0804.

29 *Pseudacris crucifer* (spring peeper): 0222, 0262, 0276, 0280, 0362, 0643, 0658, 1051.
30 *Pseudacris nigrita* (southern chorus frog): 0222, 0280, 0643, 1051, 1118.
31 *Pseudacris ocularis* (little grass frog): 0222, 0362, 0643, 0804.
32 *Pseudacris ornata* (ornate chorus frog): 0222, 0280, 0362, 0643, 0804, 1051, 1118, 1301.
33 *Pseudacris triseriata* (striped chorus frog): 0222, 0643.
34 *Gastrophryne carolinensis* (eastern narrow-mouthed toad): 0222, 0280, 0362, 0643, 0658, 0804, 1301.
35 *Rana areolata* (crawfish or Carolina gopher frog): 0222, 0362, 0643, 0804.
36 *Rana catesbeiana* (bullfrog): 0209, 0222, 0272, 0279, 0362, 0643, 0658, 0804.
37 *Rana clamitans* (green or bronze frog): 0362, 0643, 0658, 0804.
38 *Rana grylio* (pig frog): 0222, 0276, 0280, 0643.
39 *Rana palustris* (pickerel frog): 0222, 0362, 0643.
40 *Rana sphenocephala* (southern leopard frog): 0222, 0276, 0280, 0362, 0643, 0658, 0804, 0863, 1008, 1051.
41 *Rana virgatipes* (carpenter frog): 0222, 0643.

ORDER CROCODILIA (Crocodilians)
42 *Alligator mississippiensis* (American alligator): 0079, 0209, 0326, 0391, 0405, 0440, 0553, 0566, 0612, 0643, 0664, 0701, 0717, 0744, 0775, 0812, 0821, 1078, 1304.

ORDER CHELONIA (Turtles)
43 *Chelydra serpentina* (common snapping turtle): 0209, 0326, 0507, 0565, 0566, 0643, 0678, 0742, 0804, 0827, 0852, 0892, 0908, 0999, 1007, 1019, 1073, 1127, 1342.
44 *Kinosternon bauri* (striped mud turtle): 0643, 0875, 0882.
45 *Sternotherus odoratus* (stinkpot): 0209, 0244, 0263, 0318, 0326, 0565, 0566, 0603, 0643, 0652, 0742, 0779, 0804, 0816, 0827, 0866, 0868, 0875, 0999, 1007, 1019, 1073, 1127, 1184.
46 *Kinosternon subrubrum* (eastern mud turtle): 0209, 0244, 0263, 0318, 0326, 0565, 0566, 0603, 0643, 0652, 0742, 0779, 0804, 0816, 0827, 0866, 0868, 0875, 0882, 0999, 1007, 1019, 1073, 1127.
47 *Pseudemys concinna* (river cooter): 0507, 0566, 0643, 0836, 0999.
48 *Pseudemys floridana* (Florida cooter): 0209, 0326, 0520, 0565, 0566, 0643, 0779, 0816, 0866, 0892, 0999, 1007, 1019, 1127.
49 *Chrysemys picta* (painted turtle): 0199, 0200, 0216, 0263, 0507, 0565, 0630, 0643, 0678, 0718, 0742, 0793, 0812, 0816, 0827, 0852, 0858, 0868, 1007, 1151, 1342.

50 *Trachemys scripta* (slider turtle): 0199, 0200, 0209, 0216, 0227, 0244, 0326, 0381, 0384, 0405, 0507, 0565, 0566, 0612, 0629, 0630, 0643, 0664, 0678, 0682, 0717, 0718, 0722, 0742, 0779, 0793, 0804, 0816, 0827, 0836, 0852, 0858, 0866, 0868, 0892, 0908, 0999, 1007, 1019, 1044, 1091, 1112, 1127, 1151, 1260, 1268, 1342.
51 *Clemmys guttata* (spotted turtle): 0565, 0566, 0643, 1282.
52 *Deirochelys reticularia* (chicken turtle): 0200, 0209, 0244, 0263, 0326, 0565, 0566, 0570, 0603, 0643, 0652, 0779, 0804, 0816, 0852, 0858, 0866, 0999, 1007, 1019, 1127, 1151.
53 *Terrapene carolina* (eastern box turtle): 0302, 0565, 0566, 0643, 0804, 0999, 1073, 1111.
54 *Trionyx spiniferus* (spiny softshell turtle): 0565, 0566, 0643.

ORDER SQUAMATA (Snakes and lizards)

SUBORDER LACERTILIA (Lizards)

55 *Anolis carolinensis* (green anole or chameleon): 0276, 0315, 0566, 0643, 0804, 0869, 0963, 1336.
56 *Sceloporus undulatus* (eastern fence lizard): 0276, 0566, 0643, 0662, 0676, 0782, 0804, 0812, 1139.
57 *Cnemidophorus sexlineatus* (six-lined racerunner): 0048, 0276, 0566, 0643, 0676, 0804.
58 *Eumeces fasciatus* (five-lined skink): 0566, 0643, 0804, 0869, 0886, 0903, 0963, 1011, 1022, 1050, 1056, 1095, 1102, 1110, 1169, 1326.
59 *Eumeces inexpectatus* (southeastern five-lined skink): 0566, 0643, 0804, 0869, 0903, 1022, 1050, 1056, 1095, 1110, 1169.
60 *Eumeces laticeps* (broadheaded skink): 0566, 0643, 0804, 0869, 0886, 0903, 0963, 0993, 1011, 1031, 1032, 1033, 1050, 1056, 1063, 1064, 1095, 1102, 1110, 1143, 1160, 1169, 1211, 1297, 1309, 1326.
61 *Scincella lateralis* (ground skink): 0643, 0804, 0886, 0903.
62 *Ophisaurus attenuatus* (slender glass lizard): 0566, 0643, 0804.
63 *Ophisaurus ventralis* (eastern glass lizard): 0566, 0643, 0804.

SUBORDER SERPENTES (Snakes)

64 *Carphophis amoenus* (worm snake): 0566.
65 *Cemophora coccinea* (scarlet snake): 0316, 0566, 0643, 0804, 1133.
66 *Coluber constrictor* (racer or black racer): 0276, 0301, 0396, 0397, 0439, 0566, 0643, 0804, 0886, 0903, 1133.
67 *Diadophis punctatus* (ringneck snake): 0566, 0643, 0742, 0804, 1133.
68 *Elaphe guttata* (corn snake): 0315, 0396, 0439, 0475, 0566, 0624, 0643.
69 *Elaphe obsoleta* (rat snake): 0396, 0397, 0439, 0566, 0643, 0710, 0804, 0983.
70 *Farancia abacura* (mud snake): 0396, 0566, 0643, 0804, 1210.

71 *Farancia erytrogramma* (rainbow snake): 0396, 0543, 0566, 0643, 0728.
72 *Heterodon platyrhinos* (eastern hognose snake): 0396, 0439, 0475, 0566, 0624, 0643, 0804, 1133.
73 *Heterodon simus* (southern hognose snake): 0396, 0566, 0643, 0804, 1133.
74 *Lampropeltis getulus* (common kingsnake): 0152, 0396, 0397, 0439, 0555, 0566, 0643, 0804, 1073, 1133.
75 *Lampropeltis triangulum* (milk snake or scarlet kingsnake): 0566, 0643.
76 *Masticophis flagellum* (coachwhip): 0396, 0397, 0439, 0566, 0643.
77 *Nerodia cyclopion* (green water snake): 0209, 0566, 0643.
78 *Nerodia erythrogaster* (red-bellied water snake): 0396, 0439, 0566, 0643, 0804.
79 *Nerodia fasciata* (banded water snake): 0566, 0624, 0643, 0804, 0815.
80 *Nerodia sipedon* (northern water snake): 0209, 0276, 0326, 0392, 0396, 0397, 0439, 0643, 0744.
81 *Nerodia taxispilota* (brown water snake): 0276, 0326, 0396, 0397, 0439, 0566, 0589, 0624, 0643, 0815, 0983.
82 *Opheodrys aestivus* (rough green snake): 0396, 0566, 0643.
83 *Pituophis melanoleucus* (pine snake): 0396, 0439, 0566, 0643, 0804.
84 *Regina rigida* (glossy crayfish snake): 0643.
85 *Regina septemvittata* (queen snake): none.
86 *Rhadinaea flavilata* (yellow-lipped snake): 0566, 1227.
87 *Seminatrix pygaea* (black swamp snake): 0566, 0643, 0804.
88 *Storeria dekayi* (brown snake): 0566, 0643, 0804, 1133.
89 *Storeria occipitomaculata* (red-bellied snake): 0566, 0643, 0804, 0885, 1133.
90 *Tantilla coronata* (southeastern crowned snake): 0566, 0643, 0728, 0804, 1133.
91 *Thamnophis sauritus* (eastern ribbon snake): 0396, 0566, 0643, 0728, 0804.
92 *Thamnophis sirtalis* (common garter snake): 0392, 0396, 0566, 0643, 0804, 0812, 0903, 1133.
93 *Virginia striatula* (rough earth snake): 0566, 0804.
94 *Virginia valeriae* (smooth earth snake): 0566, 0643, 0804, 1133.
95 *Micrurus fulvius* (eastern coral snake): 0566, 0643.
96 *Agkistrodon contortrix* (copperhead): 0396, 0439, 0566, 0643.
97 *Agkistrodon piscivorus* (cottonmouth): 0305, 0396, 0566, 0643, 0804.
98 *Crotalus horridus* (timber or canebrake rattlesnake): 0301, 0305, 0396, 0439, 0566, 0643, 0804.
99 *Sistrurus miliarius* (pygmy rattlesnake): 0566, 0643.

SREL Herpetological Reprints

Numbers preceding references are SREL reprint numbers.

0048 Bellis, E. D. 1964. A summer six-lined racerunner (*Cnemidophorus sexlineatus*) population in South Carolina. *Herpetologica* 20(1):9–16. Species: 57; Subject categories: 1, 2, 3, 5, 6, 7, 9, 10, 22.

0079 Jenkins, J. H., and E. E. Provost. 1964. *The population status of the larger vertebrates on the Atomic Energy Commission Savannah River Plant Site.* U.S. Atomic Energy Commission, TID-19562, p. 145. Species: 42; Subject categories: 2, 7.

0152 Brisbin, I. L., Jr. 1968. Evidence for the use of postanal musk as an alarm device in the king snake, *Lampropeltis getulus*. *Herpetologica* 24(2):169–170. Species: 74; Subject categories: 2, 10.

0199 Clark, D. B., and J. W. Gibbons. 1969. Dietary shift in the turtle *Pseudemys scripta* (Schoepff) from youth to maturity. *Copeia* 1969(4):704–706. Species: 49, 50; Subject categories: 1, 2, 3, 4, 12, 22.

0200 Gibbons, J. W. 1969. Ecology and population dynamics of the chicken turtle, *Deirochelys reticularia*. *Copeia* 1969(4):669–676. Species: 49, 50, 52; Subject categories: 1, 2, 3, 5, 6, 9, 10, 22.

0209 Gibbons, J. W. 1970. Terrestrial activity and the population dynamics of aquatic turtles. *Am. Midl. Nat.* 83(2):404–414. Species: 36, 42, 43, 45, 46, 48, 50, 52, 77, 80; Subject categories: 1, 2, 5, 7, 9, 10, 22.

0216 Gibbons, J. W. 1970. Reproductive dynamics of a turtle (*Pseudemys scripta*) population in a reservoir receiving heated effluent from a nuclear reactor. *Can. J. Zool.* 48(4):881–885. Species: 49, 50; Subject categories: 1, 2, 3, 5, 6, 7, 22.

0222 Altig, R. 1970. A key to the tadpoles of the continental United States and Canada. *Herpetologica* 26(2):180–207. Species: 17, 18, 19, 20, 21, 22, 23, 24, 25, 26, 27, 28, 29, 30, 31, 32, 33, 34, 35, 36, 38, 39, 40, 41; Subject categories: 2, 14, 15, 16, 21.

0227 Schmidt, G. D., G. W. Esch, and J. W. Gibbons. 1970. *Neoechinorhynchus chelonos*, a new species of Acanthocephalan parasite of turtles. *Proc. Helminthological Soc. Washington* 37(2):172–174. Species: 50; Subject categories: 2, 13.

0244 Bennett, D. H., J. W. Gibbons, and J. C. Franson. 1970. Terrestrial activity in aquatic turtles. *Ecology* 51(4):738–740. Species: 45, 46, 50, 52; Subject categories: 1, 2, 9, 10, 18, 22.

0262 Farrell, M. P. 1971. Effect of temperature and photoperiod acclimations on the water economy of *Hyla crucifer*. *Herpetologica* 27(1):41–48. Species: 26, 29; Subject categories: 2, 4, 11, 22.

0263 Gibbons, J. W. 1970. Sex ratios in turtles. *Res. Popul. Ecol.* 7:252–254. Species: 45, 46, 49, 52; Subject categories: 22.

0272 Goodyear, C. P., and R. Altig. 1971. Orientation of bullfrogs, *Rana catesbeiana* during metamorphosis. *Copeia* 1971(2):362–364. Species: 36; Subject categories: 2, 9, 10, 22.

0276 Boyd, C. E., and C. P. Goodyear. 1971. The protein content of some common reptiles and amphibians. *Herpetologica* 27(3):317–320. Species: 7, 8, 12, 17, 19, 21, 24, 25, 26, 28, 29, 38, 40, 55, 56, 57, 66, 80, 81; Subject categories: 4, 22.

0279 Goodyear, C. P. 1971. Y-axis orientation of the oak toad, *Bufo quercicus*. *Herpetologica* 27(3):320–323. Species: 18, 21, 36; Subject categories: 1, 2, 9, 10, 22.

0280 Boyd, C. E., and C. P. Goodyear. 1971. Somatic and gametic dry matter and protein in gravid females of several amphibian species. *Comp. Biochem. Physiol.* 40A:771–775. Species: 17, 18, 19, 21, 27, 28, 29, 30, 32, 34, 38, 40; Subject categories: 4, 5, 11, 12.

0301 Cale, W. G., Jr., and J. W. Gibbons. 1972. Relationships between body size, size of the fat bodies, and total lipid content in the canebrake rattlesnake (*Crotalus horridus*) and the black racer (*Coluber constrictor*). *Herpetologica* 28(1):51–53. Species: 66, 98; Subject categories: 2, 4, 22.

0302 Brisbin, I. L., Jr. 1972. Seasonal variations in the live weights and major body components of captive box turtles. *Herpetologica* 28(1):70–75. Species: 53; Subject categories: 4, 22.

0305 Gibbons, J. W. 1972. Reproduction, growth, and sexual dimorphism in the canebrake rattlesnake (*Crotalus horridus atricaudatus*). *Copeia* 1972(2):222–226. Species: 97, 98; Subject categories: 1, 2, 3, 4, 5, 6, 22.

0315 Smith, G. C., and D. Watson. 1972. Selection patterns of corn snakes, *Elaphe guttata*, of different phenotypes of the house mouse, *Mus musculus*. *Copeia* 1972(3):529–532. Species: 55, 68; Subject categories: 2, 10, 22.

0316 Nelson, D. H., and J. W. Gibbons. 1972. Ecology, abundance, and seasonal activity of the scarlet snake, *Cemophora coccinea*. *Copeia* 1972(3):582–584. Species: 65; Subject categories: 1, 2, 7, 9, 10, 22.

0318 Bennett, D. H. 1972. Notes on the terrestrial wintering of mud turtles (*Kinosternon subrubrum*). *Herpetologica* 28(3):245–247. Species: 45, 46; Subject categories: 1, 2, 9, 10, 18, 22.

0326 Parker, E. D., M. F. Hirshfield, and J. W. Gibbons. 1973. Ecological comparisons of thermally affected aquatic environments. *J. Water Pollution Control Fed.* 45(4):726–733. Species: 42, 43, 45, 46, 48, 50, 52, 80, 81; Subject categories: 1, 2.

0362 Gibbons, J. W., and D. H. Bennett. 1974. Determination of anuran terrestrial activity patterns by a drift fence method. *Copeia* 1974(1):236–243. Species: 17, 18, 19, 21, 25, 26, 28, 29, 31, 32, 34, 35, 36, 37, 39, 40; Subject categories: 1, 2, 7, 9, 22.

0381 Bourque, J. E., and G. W. Esch. 1974. Population ecology of parasites in turtles from thermally altered and natural aquatic communities. In *Thermal ecology*, ed. J. W. Gibbons and R. R. Sharitz, pp. 551–561 (CONF-730505), Technical Information Center, Office of Information Services, U.S. Atomic Energy Commission, Oak Ridge, Tennessee. Species: 50; Subject categories: 1, 2, 13.

0384 Christy, E. J., J. O. Farlow, J. E. Bourque, and J. W. Gibbons. 1974. Enhanced growth and increased body size of turtles living in thermal and post-thermal aquatic systems. In *Thermal ecology*, ed. J. W. Gibbons and R. R. Sharitz, pp. 277–284 (CONF-730505), Technical Information Center, Office of Information Services, U.S. Atomic Energy Commission, Oak Ridge, Tennessee. Species: 50; Subject categories: 1, 2, 3, 12, 22.

0391 Murphy, T. M., Jr., and I. L. Brisbin, Jr. 1974. Distribution of alligators in response to thermal gradients in a reactor cooling reservoir. In *Thermal ecology*, ed. J. W. Gibbons and R. R. Sharitz, pp. 313–321 (CONF-730505), Technical Information Center, Office of Information Services, U.S. Atomic Energy Commission, Oak Ridge, Tennessee. Species: 42; Subject categories: 1, 2, 7, 9, 22.

0392 Nelson, D. 1974. Growth and developmental responses of larval toad populations to heated effluent in a South Carolina reservoir. In *Thermal ecology*, ed. J. W. Gibbons and R. R. Sharitz, pp. 264–276 (CONF-730505), Technical Information Center, Office of Information Services, U.S. Atomic Energy Commission, Oak Ridge, Tennessee. Species: 19, 80, 92; Subject categories: 1, 2, 3, 6, 22.

0396 Brisbin, I. L., Jr., M., A. Staton, J. E. Pinder III, and R. A. Geiger. 1974. Radiocesium concentrations of snakes from contaminated and non-contaminated habitats of the AEC Savannah River Plant. *Copeia* 1974(2):501–506. Species: 66, 68, 69, 70, 71, 72, 73, 74, 76, 78, 80, 81, 82, 83, 91, 92, 96, 97, 98; Subject categories: 1, 2, 18, 22.

0397 Staton, M. A., I. L. Brisbin, Jr., and R. A. Geiger. 1974. Some aspects of radiocesium retention in naturally contaminated captive snakes. *Herpetologica* 30(2):204–211. Species: 66, 69, 74, 76, 80, 81; Subject categories: 2, 4, 18, 22.

0405 Gibbons, J. W., and R. R. Sharitz. 1974. Thermal alteration of aquatic ecosystems. *Am. Sci.* 62(6):660–670. Species: 42, 50; Subject categories: 1, 2, 9, 13.

0429 Dapson, R. W., and L. Kaplan. 1975. Biological half-life and distribution of radiocesium in a contaminated population of green treefrogs *Hyla cinerea*. *Oikos* 26:39–42. Species: 25; Subject categories: 1, 2, 18, 22.

0439 Kaufman, G. A., and J. W. Gibbons. 1975. Weight-length relationships in thirteen species of snakes in the southeastern United States. *Herpetologica* 31(1):31–37. Species: 66, 68, 69, 72, 74, 76, 78, 80, 81, 83, 96, 98; Subject categories: 2, 14, 22.

0440 Murphy, T. M., Jr., and T. T. Fendley. 1974. A new technique for live trapping of nuisance alligators. In *Proc. 27th Ann. Conf. Southeastern Assoc. Game and Fish Commissioners* 27:308-322. Columbia, SC. Species: 42; Subject categories: 22.

0475 Smith, G. C. 1976. Ecological energetics of three species of ectothermic vertebrates. *Ecology* 57(2):252–264. Species: 19, 68, 72; Subject categories: 2, 3, 4, 5, 11, 12, 18, 22.

0507 Gibbons, J. W. 1976. Aging phenomena in reptiles. In *Special review of experimental aging research*, ed. M. F. Elias, B. E. Eleftheriou, and P. K. Elias, pp. 453–475, EAR, Inc., Bar Harbor, Maine. Species: 43, 47, 49, 50; Subject categories: 3, 4, 6, 11.

0520 Gibbons, J. W., and J. W. Coker. 1977. Ecological and life history aspects of the cooter, *Chrysemys floridana* (Le Conte). *Herpetologica* 33(1):29–33. Species: 48; Subject categories: 1, 2, 3, 5, 6, 9, 22.

0543 Gibbons, J. W., J. W. Coker, and T. M. Murphy, Jr. 1977. Selected aspects of the life history of the rainbow snake (*Farancia erytrogramma*). *Herpetologica* 33(3):276–281. Species: 71; Subject categories: 1, 2, 5, 6, 9, 14, 22.

0553 Standora, E. A. 1977. An eight-channel radio telemetry system to monitor alligator body temperatures in a heated reservoir. In Proc. First Int. Conf. Wildlife Biotelemetry, ed. F. M. Long, pp. 70–78, Laramie, Wyoming. Species: 42; Subject categories: 2, 20, 22.

0555 Williams, P. R., Jr., and I. L. Brisbin, Jr. 1978. Responses of captive-reared eastern kingsnakes (*Lampropeltis getulus*) to several prey odor stimuli. *Herpetologica* 34(1):79–83. Species: 74; Subject categories: 10, 22.

0565 Gibbons, J. W., and D. H. Nelson. 1978. The evolutionary significance of delayed emergence from the nest by hatchling turtles. *Evolution* 32(2):297–303. Species: 43, 45, 46, 48, 49, 50, 51, 52, 53, 54; Subject categories: 1, 2, 3, 5, 9, 10, 19, 22.

0566 Gibbons, J. W. 1978. Reptiles. In *Annotated checklist of the biota of the coastal zone of South Carolina*, ed. R. G. Zingmark, pp. 270–276. University of South Carolina Press, Columbia. Species: 42, 43, 45, 46,

47, 48, 50, 51, 52, 53, 54, 55, 56, 57, 58, 59, 60, 62, 63, 64, 65, 66, 67, 68, 69, 70, 71, 72, 73, 74, 75, 76, 77, 78, 79, 81, 82, 83, 86, 87, 88, 89, 90, 91, 92, 93, 94, 95, 96, 97, 98, 99; Subject categories: 1, 2.

0570 Gibbons, J. W., and J. L. Greene. 1978. Selected aspects of the ecology of the chicken turtle, *Deirochelys reticularia* (Latreille) (Reptilia, Testudines, Emydidae). *J. Herpetol.* 12(2):237–241. Species: 52; Subject categories: 1, 2, 3, 5, 9, 22.

0589 Semlitsch, R. D., and J. W. Gibbons. 1978. Reproductive allocation in the brown water snake, *Natrix taxispilota*. *Copeia* 1978(4):721–723. Species: 81; Subject categories: 1, 2, 4, 5, 22.

0590 Patterson, K. K. 1978. Life history aspects of paedogenic populations of the mole salamander, *Ambystoma talpoideum*. *Copeia* 1978(4): 649–655. Species: 6, 7, 8; Subject categories: 1, 2, 3, 5, 7, 9, 14, 22.

0603 Gibbons, J. W., and J. L. Greene. 1979. X-ray photography: A technique to determine reproductive patterns of freshwater turtles. *Herpetologica* 35(1):86–89. Species: 45, 46, 52; Subject categories: 5, 18, 22.

0612 Gibbons, J. W., G. H. Keaton, J. P. Schubauer, J. L. Greene, D. H. Bennett, J. R. McAuliffe, and R. R. Sharitz. 1979. Unusual population size structure in freshwater turtles on barrier islands. *Georgia J. Sci.* 37:155–159. Species: 42, 50; Subject categories: 1, 2, 3, 5, 22.

0624 Semlitsch, R. D. 1979. The influence of temperature on ecdysis rates in snakes (genus *Natrix*) (Reptilia, Serpentes, Colubridae). *J. Herpetol.* 13(2):212–214. Species: 68, 72, 79, 81; Subject categories: 1, 2, 11, 12, 22.

0629 Esch, G. W., J. W. Gibbons, and J. E. Bourque. 1979. Species diversity of helminth parasites in *Chrysemys s. scripta* from a variety of habitats in South Carolina. *J. Parasitol.* 65(4):633–638. Species: 50; Subject categories: 1, 2, 13.

0630 Esch, G. W., J. W. Gibbons, and J. E. Bourque. 1979. The distribution and abundance of enteric helminths in *Chrysemys s. scripta* from various habitats on the Savannah River Plant in South Carolina. *J. Parasitol.* 65(4):624–632. Species: 49, 50; Subject categories: 1, 2, 13.

0643 Gibbons, J. W., D. H. Nelson, K. K. Patterson, and J. L. Greene. 1976. The reptiles and amphibians of the Savannah River Plant in west-central South Carolina. In *Proc. First South Carolina Endangered Species Symp.*, ed. D. N. Forsythe and W. B. Ezell, Jr., pp. 133–143. Species: 1, 2, 3, 4, 5, 6, 7, 8, 9, 10, 11, 12, 13, 14, 15, 16, 17, 18, 19, 20, 21, 22, 23, 24, 25, 26, 27, 28, 29, 30, 31, 32, 33, 34, 35, 36, 37, 38, 39, 40, 41, 42, 43, 44, 45, 46, 47, 48, 49, 50, 51, 52, 53, 54, 55, 56, 57, 58, 59, 60,

61, 62, 63, 65, 66, 67, 68, 69, 70, 71, 72, 73, 74, 75, 76, 77, 78, 79, 80, 81, 82, 83, 84, 87, 88, 89, 90, 91, 92, 94, 95, 96, 97, 98, 99; Subject categories: 1, 2, 7, 22.

0652 Gibbons, J. W., J. L. Greene, and J. P. Schubauer. 1978. Variability in clutch size in aquatic chelonians. *British J. Herpetol.* 6:13–14. Species: 45, 46, 52; Subject categories: 5, 18.

0658 Bennett, S. H., J. W. Gibbons, and J. Glanville. 1980. Terrestrial activity, abundance and diversity of amphibians in differently managed forest types. *Am. Midl. Nat.* 103(2):412–416. Species: 6, 7, 9, 13, 14, 17, 18, 19, 24, 25, 28, 29, 34, 36, 37, 40; Subject categories: 1, 2, 7, 9, 17, 20, 22.

0662 Ferguson, G. W., and T. Brockman. 1980. Geographic differences of growth rate of *Sceloporus* lizards (Sauria: Iguanidae). *Copeia* 1980(2):259–264. Species: 56; Subject categories: 2, 3, 4, 5, 11, 22.

0664 Gibbons, J. W., R. R. Sharitz, and I. L. Brisbin, Jr. 1980. Thermal ecology research at the Savannah River Plant: A review. *Nuclear Safety* 21(3):367–379. Species: 19, 42, 50; Subject categories: 1, 2, 3, 5, 8, 9, 12, 17, 19, 20.

0666 Semlitsch, R. D. 1980. Geographic and local variation in population parameters of the slimy salamander *Plethodon glutinosus*. *Herpetologica* 36(1):6–16. Species: 14; Subject categories: 1, 2, 3, 5, 6, 7, 22.

0675 Semlitsch, R. D., and M. A. McMillan. 1980. Breeding migrations, population size structure, and reproduction of the dwarf salamander, *Eurycea quadridigitata*, in South Carolina. *Brimleyana* 3:97–105. Species: 7, 13; Subject categories: 1, 2, 5, 6, 7, 9, 22.

0676 Ferguson, G. W., C. H. Bohlen, and H. P. Woolley. 1980. *Sceloporus undulatus*: Comparative life history and regulation of a Kansas population. *Ecology* 61(2):313–322. Species: 56, 57; Subject categories: 1, 2, 5, 6.

0677 Semlitsch, R. D. 1980. Growth and metamorphosis of larval dwarf salamanders (*Eurycea quadridigitata*). *Herpetologica* 36(2):138–140. Species: 11, 13; Subject categories: 1, 2, 3, 5, 6, 9, 22.

0678 Parmenter, R. R. 1980. Effects of food availability and water temperature on the feeding ecology of pond sliders (*Chrysemys s. scripta*). *Copeia* 1980(3):503–514. Species: 43, 49, 50; Subject categories: 1, 2, 11, 22.

0682 Gibbons, W., and J. Caldwell. 1980. Herpetology at the Savannah River Ecology Laboratory. *Herp. Rev.* 11(3):72–74. Species: 50; Subject categories: 2.

0684 Caldwell, J. P., J. H. Thorp, and T. O. Jervy. 1980. Predator-prey relationships among larval dragonflies, salamanders, and frogs.

Oecologia (Berlin) 46:285–289. Species: 7, 28; Subject categories: 1, 2, 3, 6, 7, 9, 22.

0701 Adams, S. E., M. H. Smith, and R. Baccus. 1980. Biochemical variation in the American alligator. *Herpetologica* 36(4):289–296. Species: 42; Subject categories: 2, 8, 22.

0706 McMillan, M. A., and R. D. Semlitsch. 1980. Prey of the dwarf salamander, *Eurycea quadridigitata*, in South Carolina. *J. Herpetol.* 14(4):424–426. Species: 13; Subject categories: 1, 2, 11, 22.

0710 Fendley, T. T. 1980. Incubating wood duck and hooded merganser hens killed by black rat snakes. *Wilson Bull.* 92(4):526–527. Species: 69; Subject categories: 1, 10, 11, 20.

0716 Semlitsch, R. D. 1981. Terrestrial activity and summer home range of the mole salamander (*Ambystoma talpoideum*). *Can. J. Zool.* 59:315–322. Species: 5, 7, 8; Subject categories: 1, 2, 6, 9, 10, 18, 22.

0717 Gibbons, J. W., and R. R. Sharitz. 1981. Thermal ecology: Environmental teachings of a nuclear reactor site. *BioScience*, 31(4):293–298. Species: 42, 50; Subject categories: 1.

0718 Gibbons, J. W., R. D. Semlitsch, J. L. Greene, and J. P. Schubauer. 1981. Variation in age and size at maturity of the slider turtle (*Pseudemys scripta*). *Am. Nat.* 117:841–845. Species: 49, 50; Subject categories: 2, 3, 5, 14.

0722 Schubauer, J. P. 1981. A reliable radio-telemetry tracking system suitable for studies of chelonians. *J. Herpetol.* 15(1):117–120. Species: 50; Subject categories: 17, 20, 22.

0728 Semlitsch, R. D., K. L. Brown, and J. P. Caldwell. 1981. Habitat utilization, seasonal activity, and population size structure of the southeastern crowned snake *Tantilla coronata*. *Herpetologica* 37(1): 40–46. Species: 71, 90, 91; Subject categories: 1, 2, 5, 6, 7, 10, 14, 22.

0742 Parmenter, R. R. 1981. Digestive turnover rates in freshwater turtles: The influence of temperature and body size. *Comp. Biochem. Physiol.* 70A:235–238. Species: 43, 45, 46, 49, 50, 67; Subject categories: 10, 11, 12, 22.

0743 Semlitsch, R. D. 1981. Effects of implanted tantalum-182 wire tags on the mole salamander, *Ambystoma talpoideum*. *Copeia* 1981(3): 735–737. Species: 7, 11; Subject categories: 2, 17, 18, 20, 22.

0744 Murphy, P. A. 1981. Celestial compass orientation in juvenile American alligators (*Alligator mississippiensis*). *Copeia* 1981(3):638–645. Species: 42, 80; Subject categories: 2, 9, 11, 22.

0775 Brisbin, I. L., Jr., E. A. Standora, and M. J. Vargo. 1982. Body temperatures and behavior of American alligators during cold winter weather. *Am. Midl. Nat.* 107(2):209–218. Species: 42; Subject categories: 1, 2, 6, 10, 22.

0779 Gibbons, J. W. 1982. Reproductive patterns in freshwater turtles. *Herpetologica* 38(1):222–227. Species: 45, 46, 48, 50, 52; Subject categories: 1, 2, 5, 22.

0782 Vitt, L. J., and H. J. Price. 1982. Ecological and evolutionary determinants of relative clutch mass in lizards. *Herpetologica* 38(1):237–255. Species: 56; Subject categories: 5.

0792 Semlitsch, R. D., and J. P. Caldwell. 1982. Effects of density on growth, metamorphosis, and survivorship in tadpoles of *Scaphiopus holbrooki*. *Ecology* 63(4):905–911. Species: 17; Subject categories: 3, 22.

0793 Gibbons, J. W., and R. D. Semlitsch. 1982. Survivorship and longevity of a long-lived vertebrate species: How long do turtles live? *J. Anim. Ecol.* 51:523–527. Species: 49, 50; Subject categories: 1, 2, 6, 22.

0804 Gibbons, J. W., and R. D. Semlitsch. 1981. Terrestrial drift fences with pitfall traps: An effective technique for quantitative sampling of animal populations. *Brimleyana* 7:1–16. Species: 6, 7, 8, 9, 11, 13, 14, 15, 16, 17, 18, 19, 21, 23, 25, 26, 27, 28, 31, 32, 34, 35, 36, 37, 40, 43, 45, 46, 50, 52, 53, 55, 56, 57, 58, 59, 60, 61, 62, 63, 65, 66, 67, 69, 70, 72, 73, 74, 78, 79, 83, 87, 88, 89, 90, 91, 92, 93, 94, 97, 98; Subject categories: 2, 7, 9, 10, 17, 20, 22.

0812 Congdon, J. D., A. E. Dunham, and D. W. Tinkle. 1982. Energy budgets and life histories of reptiles. In *Biology of the Reptilia*, ed. C. Gans, 13:233–271. Species: 42, 49, 56, 92; Subject categories: 5, 12, 21.

0815 Semlitsch, R. D., and J. W. Gibbons. 1982. Body size dimorphism and sexual selection in two species of water snakes. *Copeia* 1982(4):974–976. Species: 79, 81; Subject categories: 5, 19, 22.

0816 Gibbons, J. W., J. L. Greene, and K. K. Patterson. 1982. Variation in reproductive characteristics of aquatic turtles. *Copeia* 1982(4):776–784. Species: 45, 46, 48, 49, 50, 52; Subject categories: 2, 5, 22.

0821 Brisbin, I. L., Jr. 1982. Applied ecological studies of the American alligator at the Savannah River Ecology Laboratory: An overview of program goals and design. In *Proc. Fifth Working Meeting of the Crocodile Specialist Group, SSC/IUCN*, pp. 376–388, Gland, Switzerland. Species: 42; Subject categories: 21.

0827 Congdon, J. D., D. W. Tinkle, and P. C. Rosen. 1983. Egg components and utilization during development in aquatic turtles. *Copeia* 1983(1):264–268. Species: 43, 45, 46, 49, 50; Subject categories: 4, 5, 22.

0836 Schubauer, J. P., and R. R. Parmenter. 1981. Winter feeding by aquatic turtles in a southeastern reservoir. *J. Herpetol.* 15(4):444–447. Species: 47, 50; Subject categories: 1, 2, 10, 11, 22.

0840 Semlitsch, R. D. 1983. Growth and metamorphosis of larval red salamanders (*Pseudotriton ruber*) on the coastal plain of South Carolina. *Herpetologica* 39(1):48–52. Species: 2, 3, 10, 11, 12, 15, 16; Subject categories: 1, 2, 3, 6, 22.

0844 Semlitsch, R. D. 1983. Burrowing ability and behavior of salamanders of the genus *Ambystoma. Can. J. Zool.* 61:616–620. Species: 5, 6, 7, 8; Subject categories: 10, 14, 22.

0848 Semlitsch, R. D., and C. A. West. 1983. Aspects of the life history and ecology of Webster's salamander, *Plethodon websteri. Copeia* 1983(2):339–346. Species: 9, 10, 11, 12, 14; Subject categories: 1, 3, 5, 6.

0852 Congdon, J. D., J. W. Gibbons, and J. L. Greene. 1983. Parental investment in the chicken turtle (*Deirochelys reticularia*). *Ecology* 64(3):419–425. Species: 43, 49, 50, 52; Subject categories: 3, 4, 5, 15, 22.

0858 Congdon, J. D., and J. W. Gibbons. 1983. Relationships of reproductive characteristics to body size in *Pseudemys scripta. Herpetologica* 39(2):147–151. Species: 49, 50, 52; Subject categories: 1, 2, 5, 22.

0859 Vitt, L. J. 1983. Tail loss in lizards: The significance of foraging and predator escape modes. *Herpetologica* 39(2):151–162.

0863 Semlitsch, R. D. 1983. Structure and dynamics of two breeding populations of the eastern tiger salamander, *Ambystoma tigrinum. Copeia* 1983(3):608–616. Species: 5, 8, 40; Subject categories: 1, 2, 5, 6, 7, 9, 22.

0866 Gibbons, J. W., J. L. Greene, and J. D. Congdon. 1983. Drought-related responses of aquatic turtle populations. *J. Herpetol.* 17(3):242–246. Species: 45, 46, 48, 50, 52; Subject categories: 1, 2, 5, 7, 9, 10, 22.

0868 Gibbons, J. W. 1983. Reproductive characteristics and ecology of the mud turtle, *Kinosternon subrubrum. Herpetologica* 39(3):254–271. Species: 45, 46, 49, 50; Subject categories: 1, 2, 3, 5, 6, 7, 9, 22.

0869 Cooper, W. E., Jr., L. J. Vitt, L. D. Vangilder, and J. W. Gibbons. 1983. Natural nest sites and brooding behavior of *Eumeces fasciatus. Herp. Rev.* 14(3):65–66. Species: 55, 58, 59, 60; Subject categories: 1, 2, 5, 10.

0875 Lamb, T. 1983. On the problematic identification of *Kinosternon* (Testudines: Kinosternidae) in Georgia, with new state localities for *Kinosternon bauri. Georgia J. Sci.* 41:115–120. Species: 44, 45, 46; Subject categories: 2, 14, 16.

0882 Lamb, T. 1983. The striped mud turtle (*Kinosternon bauri*) in South Carolina, a confirmation through multivariate character analysis. *Herpetologica* 39(4):383–390. Species: 44, 46; Subject categories: 2, 14, 15, 16, 22.

0885 Semlitsch, R. D., and G. B. Moran. 1984. Ecology of the redbelly snake (*Storeria occipitomaculata*) using mesic habitats in South Carolina. *Am. Midl. Nat.* 111(1):33–40. Species: 89; Subject categories: 1, 2, 3, 5, 6, 9, 11, 22.

0886 Cooper, W. E., and L. J. Vitt. 1984. Detection of conspecific odors by the female broad-headed skink, *Eumeces laticeps*. *J. Exp. Zool.* 229:49–54. Species: 58, 60, 61, 66; Subject categories: 2, 10, 11, 22.

0892 Avery, H. W., and L. J. Vitt. 1984. How to get blood from a turtle. *Copeia* 1984(1):209–210. Species: 43, 48, 50; Subject categories: 22.

0903 Cooper, W. E., Jr., and L. J. Vitt. 1984. Conspecific odor detection by the male broad-headed skink, *Eumeces laticeps*: Effects of sex and site of odor source and of male reproductive condition. *J. Exp. Zool.* 230:199–209. Species: 58, 59, 60, 61, 66, 92; Subject categories: 2, 10, 11, 22.

0908 Morreale, S. J., J. W. Gibbons, and J. D. Congdon. 1984. Significance of activity and movement in the yellow-bellied slider turtle (*Pseudemys scripta*). *Can. J. Zool.* 62:1038–1042. Species: 43, 50; Subject categories: 2, 5, 9.

0963 Vitt, L. J., and W. E. Cooper, Jr. 1985. The evolution of sexual dimorphism in the skink *Eumeces laticeps*: An example of sexual selection. *Can. J. Zool.* 63:995–1002. Species: 55, 58, 60; Subject categories: 1, 2, 5, 6, 11, 14, 22.

0964 Semlitsch, R. D. 1983. Terrestrial movements of an eastern tiger salamander, *Ambystoma tigrinum*. *Herp. Rev.* 14(4):112–113. Species: 8; Subject categories: 9.

0972 Semlitsch, R. D., and J. W. Gibbons. 1985. Phenotypic variation in metamorphosis and paedomorphosis in the salamander *Ambystoma talpoideum*. *Ecology* 66:1123–1130. Species: 7; Subject categories: 1, 3, 11.

0983 Bagshaw, C., and I. L. Brisbin, Jr. 1984. Long-term declines in radiocesium of two sympatric snake populations. *J. Appl. Ecol.* 21:407–413. Species: 69, 81; Subject categories: 18.

0993 Vitt, L. J., and W. E. Cooper, Jr. 1985. The relationship between reproduction and lipid cycling in the skink *Eumeces laticeps* with comments on brooding ecology. *Herpetologica* 41(4):419–432. Species: 60; Subject categories: 5, 11.

0999 Lamb, T., and J. D. Congdon. 1985. Ash content: Relationships to flexible and rigid eggshell types of turtles. *J. Herpetol.* 19:527–530. Species: 43, 45, 46, 47, 48, 50, 52, 53; Subject categories: 4, 11.

1007 Congdon, J. D., J. L. Greene, and J. W. Gibbons. 1986. Biomass of freshwater turtles: A geographic comparison. *Am. Midl. Nat.* 115: 165–173. Species: 43, 45, 46, 48, 49, 50, 52; Subject categories: 1, 7, 12.

1008 Caldwell, J. P. 1986. Selection of egg deposition sites: A seasonal shift in the southern leopard frog, *Rana sphenocephala*. *Copeia* 1986(1):249–253. Species: 40; Subject categories: 5, 10.

1011 Cooper, W. E., Jr., and L. J. Vitt. 1985. Blue tails and autotomy: Enhancement of predation avoidance in juvenile skinks. *Z. Tierpsychol.* 70:265–276. Species: 58, 60; Subject categories: 10, 14.

1019 Gibbons, J. W. 1986. Movement patterns among turtle populations: Applicability to management of the desert tortoise. *Herpetologica* 42:104–113. Species: 43, 45, 46, 48, 50, 52; Subject categories: 9.

1022 Vitt, L. J., and W. E. Cooper, Jr. 1986. Skink reproduction and sexual dimorphism: *Eumeces fasciatus* in the southeastern United States, with notes on *Eumeces inexpectatus*. *J. Herpetol.* 20:65–76. Species: 58, 59; Subject categories: 5, 14.

1031 Cooper, W. E., and L. J. Vitt. 1986. Thermal dependence of tongue-flicking and comments on use of tongue-flicking as an index of squamate behavior. *Ethology* 71:177–186. Species: 60; Subject categories: 10.

1032 Cooper, W. E., and L. J. Vitt. 1986. Tracking of female conspecific odor trails by male broad-headed skinks (*Eumeces laticeps*). *Ethology* 71:242–248. Species: 60; Subject categories: 10.

1033 Cooper, W. E., M. T. Mendonca, and L. J. Vitt. 1986. Induction of sexual receptivity in the female broad-headed skink, *Eumeces laticeps*, by estradiol-17β. *Hormones and Behavior* 10:235–242. Species: 60; Subject categories: 10, 11.

1044 Scribner, K. T., J. E. Evans, S. J. Morreale, M. H. Smith, and J. W. Gibbons. 1986. Genetic divergence among populations of the yellow-bellied slider turtle (*Pseudemys scripta*) separated by aquatic and terrestrial habitats. *Copeia* 1986(3):691–700. Species: 50; Subject categories: 1, 8.

1050 Vitt, L. J., and W. E. Cooper. 1986. Tail loss, tail color, and predator escape in *Eumeces* (Lacertilia: Scincidae): Age-specific differences in costs and benefits. *Can. J. Zool.* 64:583–592. Species: 58, 59, 60; Subject categories: 3, 6, 10, 12.

1051 Pechmann, J. H. K., and R. D. Semlitsch. 1986. Diel activity patterns in the breeding migrations of winter-breeding anurans. *Can. J. Zool.* 64:1116–1120. Species: 29, 30, 32, 40; Subject categories: 9, 10.

1056 Cooper, W. E., Jr., and L. J. Vitt. 1986. Interspecific odour discriminations among syntopic congeners in scincid lizards (genus *Eumeces*). *Behaviour* 97:1–9. Species: 58, 59, 60; Subject categories: 10, 11.

1063 Vitt, L. J., and W. E. Cooper. 1986. Foraging and diet of a diurnal predator (*Eumeces laticeps*) feeding on hidden prey. *J. Herpetol.* 20(3):408–415. Species: 60; Subject categories: 10, 12.

1064 Cooper, W. E., Jr., W. R. Garstka, and L. J. Vitt. 1986. Female sex pheromone in the lizard *Eumeces laticeps*. *Herpetologica* 42:361–366. Species: 60; Subject categories: 10, 11.

1073 Knight, J. L., and R. K. Loraine. 1986. Notes on turtle egg predation by *Lampropeltis getulus* (Linnaeus) (Reptilia: Colubridae) on the Savannah River Plant, South Carolina. *Brimleyana* 12:1–4. Species: 43, 45, 46, 53, 74; Subject categories: 10, 12.

1078 Novak, S. S., and R. A. Seigel. 1986. Gram-negative septicemia in American alligators (*Alligator mississippiensis*). *J. Wildl. Diseases* 22:484–487. Species: 42; Subject categories: 11, 20.

1091 Morreale, S. J., and J. W. Gibbons. 1986. Habitat suitability index models: Slider turtle. *U.S. Fish Wildl. Serv. Biol. Rep.* 82(10.125). Species: 50; Subject categories: 1, 2, 20.

1095 Cooper, W. E., Jr., and L. J. Vitt. 1986. Lizard pheromones: Behavioral responses and adaptive significance in skinks of the genus *Eumeces*. In *Chemical signals in vertebrates 4*, ed. D. Duvall, D. Muller-Schwarze, and R. M. Silverstein, pp. 323–340. Plenum Press, New York. Species: 58, 59, 60; Subject categories: 10, 11, 19.

1102 Cooper, W. E., Jr., and W. R. Garstka. 1987. Discrimination of male conspecific from male heterospecific odors by male scincid lizards (*Eumeces laticeps*). *J. Exp. Zoo.* 241:253–256. Species: 58, 60; Subject categories: 10, 11.

1110 Cooper, W. E., Jr., and L. J. Vitt. 1987. Intraspecific and interspecific aggression in lizards of the scincid genus *Eumeces*: Chemical detection of conspecific sexual competitors. *Herpetologica* 43:7–14. Species: 58, 59, 60; Subject categories: 10.

1111 Gatten, R. E., Jr. 1987. Cardiovascular and other physiological correlates of hibernation in aquatic and terrestrial turtles. *Am. Zool.* 27:59–68. Species: 53; Subject categories: 11.

1112 Scott, D. E., F. W. Whicker, and J. W. Gibbons. 1986. Effect of season on the retention of ^{137}Cs and ^{90}Sr by the yellow-bellied slider turtle (*Pseudemys scripta*). *Can J. Zool.* 64:2850–2853. Species: 50; Subject categories: 11, 18.

1118 Caldwell, J. P. 1987. Demography and life history of two species of chorus frogs (Anura: Hylidae) in South Carolina. *Copeia* 1987:114–127. Species: 30, 32; Subject categories: 1, 5, 6, 7, 9, 10.

1120 Semlitsch, R. D. 1987. Relationship of pond drying to the reproductive success of the salamander *Ambystoma talpoideum. Copeia* 1987:61–69. Species: 7; Subject categories: 1, 3, 5, 10.
1127 Gibbons, J. W. 1987. Why do turtles live so long? *BioScience* 37:262–269. Species: 43, 45, 46, 48, 50, 52; Subject categories: 6.
1133 Gibbons, J. W., and R. D. Semlitsch. 1987. Activity patterns. In *Snakes: Ecology and evolutionary biology*, ed. R. A. Seigel, J. T. Collins, and S. S. Novak, pp. 396–421. Macmillan, New York. Species: 65, 66, 67, 72, 73, 74, 88, 89, 90, 92, 94; Subject categories: 9.
1139 Cooper, W. E., Jr., and N. Burns. 1987. Social significance of ventrolateral coloration in the fence lizard, *Sceloporus undulatus. Anim. Behav.* 35:526–532. Species: 56; Subject categories: 10.
1143 Cooper, W. E., Jr., M. T. Mendonca, and L. J. Vitt. 1987. Induction of orange head coloration and activation of courtship and aggression by testosterone in the male broad-headed skink (*Eumeces laticeps*). *J. Herpetol.* 21:96–101. Species: 60; Subject categories: 10, 11.
1151 Congdon, J. D., and J. W. Gibbons. 1987. Morphological constraint on egg size: A challenge to optimal egg size theory? *Proc. Natl. Acad. Sci. USA* 84:4125–4147. Species: 49, 50, 52; Subject categories: 5, 15, 19.
1160 Cooper, W. E., Jr., and W. R. Garstka. 1987. Aggregation in the broad-headed skink (*Eumeces laticeps*). *Copeia* 1973(3):807–810. Species: 60; Subject categories: 10.
1162 Semlitsch, R. D. 1987. Interactions between fish and salamander larvae. *Oecologia* (Berlin) 72:481–486. Species: 5, 7; Subject categories: 10.
1164 Stangel, P. W., and R. D. Semlitsch. 1987. Experimental analysis of predation on the diel vertical migrations of a larval salamander. *Can. J. Zool.* 65:1554–1558. Species: 7; Subject categories: 10.
1169 Cooper, W. E., Jr., and L. J. Vitt. 1987. Ethological isolation, sexual behavior and pheromones in the Fasciatus species group of the lizard genus *Eumeces. Ethology* 75:328–336. Species: 58, 59, 60; Subject categories: 10, 11.
1184 Mendonca, M. T. 1987. Timing of reproductive behaviour in male musk turtles, *Sternotherus odoratus*: Effects of photoperiod, temperature and testosterone. *Anim. Behav.* 35:1002–1014. Species: 45; Subject categories: 10, 11.
1188 Meffe, G. K., and A. L. Sheldon. 1987. Habitat selection by dwarf waterdogs (*Necturus punctatus*) in South Carolina streams, with life history notes. *Herpetologica* 43:490–496. Species: 1; Subject categories: 1.

1205 Semlitsch, R. D., D. E. Scott, and J. H. K. Pechmann. 1988. Time and size at metamorphosis related to adult fitness in *Ambystoma talpoideum*. *Ecology* 69:184–192. Species: 7; Subject categories: 1, 3, 5, 7, 10.
1210 Semlitsch, R. D., J. H. K. Pechmann, and J. W. Gibbons. 1988. Annual emergence of juvenile mud snakes (*Farancia abacura*) at aquatic habitats. *Copeia* 1988(1):243–245. Species: 70; Subject categories: 1, 3, 7, 10.
1211 Cooper, W. E., Jr., and L. J. Vitt. 1988. Orange head coloration of the male broad-headed skink (*Eumeces laticeps*), a sexually selected social cue. *Copeia* 1988(1):1–6. Species: 60; Subject categories: 10.
1222 Semlitsch, R. D., and J. W. Gibbons. 1988. Fish predation in size-structured populations of treefrog tadpoles. *Oecologia* (Berlin) 75:321–326. Species: 23; Subject categories: 3, 6.
1227 Young, D. P. 1988. *Rhadinaea flavilata* (pine woods snake). *Herp. Rev.* 19(1):20. Species: 86; Subject categories: 2.
1244 Semlitsch, R. D. 1988. Allotopic distribution of two salamanders: Effects of fish predation and competitive interactions. *Copeia* 1988: 290–298. Species: 5, 7; Subject categories: 1, 2, 3, 10.
1260 Bickham, J. W., B. G. Hanks, M. J. Smolen, T. Lamb, and J. W. Gibbons. 1988. Flow cytometric analysis of the effects of low-level radiation exposure on natural populations of slider turtles (*Pseudemys scripta*). *Arch. Environ. Contam. Toxicol.* 17:837–841. Species: 50; Subject categories: 8, 18.
1268 Peters, E. L., and I. L. Brisbin, Jr. 1988. Radiocaesium elimination in the yellow-bellied turtle (*Pseudemys scripta*). *J. Appl. Ecol.* 25:461–471. Species: 50; Subject categories: 12, 18.
1280 Taylor, B. E., R. A. Estes, J. H. K. Pechmann, and R. D. Semlitsch. 1988. Trophic relations in a temporary pond: Larval salamanders and their microinvertebrate prey. *Can. J. Zool.* 66:2191–2198. Species: 6, 7, 9, 13; Subject categories: 7, 10.
1282 Lovich, J. E. 1988. Geographic variation in the seasonal activity cycle of spotted turtles, *Clemmys guttata*. *J. Herpetol.* 22(4):482–485. Species: 51; Subject categories: 10.
1297 Vitt, L. J., and W. E. Cooper, Jr. 1988. Feeding responses of skinks (*Eumeces laticeps*) to velvet ants (*Dasymutilla occidentalis*). *J. Herpetol.* 22(4):485–488. Species: 60; Subject categories: 10.
1301 Pechmann, J. H. K., D. E. Scott, J. W. Gibbons, and R. D. Semlitsch. 1989. Influence of wetland hydroperiod on diversity and abundance of metamorphosing juvenile amphibians. *Wetlands Ecol. Mgmt.* 1(1):3–11. Species: 7, 26, 32, 34; Subject categories: 1, 2, 3, 7, 9.
1304 Brisbin, I. L., Jr. 1989. Radiocesium levels in a population of Amer-

ican alligators: A model for the study of environmental contaminants in free-living crocodilians. In *Crocodiles: Proc. Eighth Working Meeting of the Crocodile Specialist Group of the Species Survival Com. Int. Union Conserv. Nature and Nat. Resources* convened at Quito, Ecuador, 13–18 October 1986. IUCN, Ave. du Mont Blanc, CH-1196, Gland, Switzerland. Species: 42; Subject categories: 18.

1307 Gibbons, J. W. 1988. The management of amphibians, reptiles and small mammals in North America: The need for an environmental attitude adjustment. Proc. Symp. Management of Amphibians, Reptiles and Small Mammals in North America, 19–21 July 1988, Flagstaff, Ariz. pp. 4–10 U.S.D.A., Forest Service, Rocky Mountain For. and Range Exper. Stat., Fort Collins, Colo. 80526, General Technical Report RM-166. Subject categories: 20.

1309 Cooper, W. E., Jr., and L. J. Vitt. 1989. Prey odor discrimination by the broad-headed skink (*Eumeces laticeps*). J. Exp. Zool. 249:11–16. Species: 60; Subject categories: 10.

1326 Vitt, L. J., and W. E. Cooper, Jr. 1989. Maternal care in skinks (*Eumeces*). *J. Herpetol.* 23(1):29–34. Species: 58, 60; Subject categories: 10.

1336 Cooper, W. E., Jr. 1989. Absence of prey odor discrimination by iguanid and agamid lizards in applicator tests. *Copeia* 1989:472–478. Species: 55; Subject categories: 10.

1342 Congdon, J. D. 1989. Proximate and evolutionary constraints on energy relations of reptiles. *Physiol. Zool.* 62(2):356–373. Species: 43, 49, 50; Subject categories: 12, 19.

Theses and Dissertations Based on Herpetological Studies on the SRS

Bourque, J. E. 1974. Studies on the population dynamics of helminth parasites in the yellow-bellied turtle, *Pseudemys scripta*. Ph.D. dissertation, Wake Forest University.

Brandt, L. A. 1989. The status and ecology of the American alligator (*Alligator mississippiensis*) in the Par Pond, Savannah River Site. M.S. thesis, Florida International University.

Caudle, A. 1984. Effects of nest humidity and egg size on hatching success and growth of embryonic freshwater chelonians. M.S. thesis, University of Georgia.

Duever, A. J. 1967. Trophic dynamics of reptiles in terms of the community food web and energy intake. M.S. thesis, University of Georgia.

Duever, M. J. 1967. Distributions in space and time of reptiles on the Savannah River Plant in South Carolina. M.S. thesis, University of Georgia.

Etges, W. J. 1979. Ecological genetic relationships in selected anurans of the southeastern United States. M.S. thesis, University of Georgia.

Hinton, T. G. 1989. The comparative kinetics of Ca, Sr, and Ra in a freshwater turtle, *Trachemys scripta*. Ph.D. thesis, Colorado State University, Fort Collins, Colorado.

Murphy, P. 1978. Celestial orientation in juvenile American alligators (*Alligator mississippiensis*). M.S. thesis, University of South Carolina.

Murphy, T. M., Jr. 1977. Distribution, movement and population dynamics of the American alligator in a thermally altered reservoir. M.S. thesis, University of Georgia.

Nelson, D. H. 1974. Ecology of anuran populations inhabiting thermally stressed aquatic ecosystems, with emphasis on larval *Rana pipiens* and *Bufo terrestris*. Ph.D. dissertation, Michigan State University.

Parmenter, Robert R. 1978. Effects of food availability and water temperature on feeding electivity, growth, and body size of pond sliders (*Chrysemys scripta* Schoepff). M.S. thesis, University of Georgia.

Patterson, K. K. 1977. The life history of the mole salamander, *Ambystoma talpoideum* (Holbrook), on the southeastern coastal plain with a possible explanation for the occurrence of paedogenesis in the species. M.S. thesis, Wake Forest University.

Peters, E. L. 1986. Radiocesium kinetics in the yellow-bellied turtle (*Pseudemys scripta*). M.S. thesis, University of Georgia.

Schubauer, J. P. 1981. The ecology and behavior of an aquatic turtle, *Pseudemys scripta*, inhabiting a thermally altered reservoir, Par Pond, South Carolina. M.A. thesis, State University, Buffalo, New York.

Semlitsch, R. D. 1984. Population ecology and reproductive strategy of the mole salamander *Ambystoma talpoideum*. Ph.D. dissertation, University of Georgia.

Smith, G. C. 1971. Ecological energetics of three ectothermic vertebrates. Ph.D. dissertation, University of Georgia.

Additional Useful References

Altig, R., and P. Ireland. 1984. A key to salamander larvae and larviform adults of the United States and Canada. *Herpetologica* 40:212–218.

Ashton, R. E., Jr., and P. S. Ashton. 1981. *Handbook of Reptiles and Amphibians of Florida*. Part one: The snakes. Windward Publishing, Miami.

———. 1985. *Handbook of reptiles and amphibians of Florida*. Part two: *Lizards, turtles and crocodilians*. Windward Publishing, Miami.

Behler, J. L., and F. W. King. 1979. *The Audubon Society field guide to North American reptiles and amphibians*. Knopf, New York.

Bennett, D. H., and R. W. McFarlane. 1983. *The fishes of the Savannah River Plant: National Environmental Research Park*. SRO-NERP-12. DOE National Environmental Research Park, Aiken, S.C.

Bishop, S. C. 1947. *Handbook of salamanders*. Comstock Publishing Company, Ithaca, N.Y.

Briese, L. A., and M. H. Smith. 1974. Seasonal abundance and movement of nine species of small mammals. *J. Mammal.* 55(3):615–629.

Carr, A. F. 1952. *Handbook of turtles; The turtles of the United States, Canada and Baja California*. Comstock Publishing Company, Ithaca, N.Y.

Cochran, D. M., and C. J. Goin. 1970. *The new field book of reptiles and amphibians*. Putnam, New York.

Collins, J. T., R. Conant, J. E. Huheey, J. L. Knight, E. M. Rundquist, and H. M. Smith. 1982. *Standard common and current scientific names for North American amphibians and reptiles*. Society for the Study of Amphibians and Reptiles. Herpetological Circular no. 12.

Conant, R. 1975. *A field guide to reptiles and amphibians of eastern and central North America*. Houghton Mifflin Company, Boston, Mass.

Duever, M. J. 1972. The striped mud turtle (*Kinosternon bauri* Garman) in South Carolina. *Herp. Rev.* 4:131.

Duke, E. K. 1984. The Savannah River Plant Environment. Prepared and printed by E. I. du Pont de Nemours and Company, SRL, Aiken, S.C., for U.S. Dept. of Energy.

Ernst, C. H., and R. W. Barbour. 1972. *Turtles of the United States*. University Press of Kentucky, Lexington.

Freeman, H. W. 1955a. Amphibians and reptiles of the SRP Area, Caudate Amphibia. University of South Carolina Publication Series III, 1:227–238.

———. 1955b. Amphibians and reptiles of the SRP Area, Chelonia. University of South Carolina Publication Series III, 1:239–244.

———. 1955c. Amphibians and reptiles of the SRP Area, Crocodilia, Sauria and Serpentes. University of South Carolina Publication Series III, 1:275–291.

———. 1956. Amphibians and reptiles of the SRP Area, Salientia. University of South Carolina Publication Series III, 2:26–35.

———. 1960. A unique environmental situation in Steed's Pond, Savannah

River Plant Area, South Carolina. University of South Carolina Publication Series III, 3:99–111.

Gibbons, J. W. 1977. *Snakes of the Savannah River Plant.* ERDA Savannah River Environmental Research Park, SRO-NERP-1, Aiken, S.C.

Gibbons, J. W., and K. K. Patterson. 1978. The reptiles and amphibians of the Savannah River Plant. DOE Savannah River National Evironmental Research Park, SRO-NERP-2, Aiken, S.C.

Golley, F. B., and J. B. Gentry. 1966. A comparison of variety and standing crop of vegetation on a one-year and a twelve-year abandoned field. *Oikos* 15:185–199.

Harrison, J. R. 1978. Amphibians. In *Annotated checklist of the biota of the coastal zone of South Carolina*, ed. R. G. Zingmark, University of South Carolina Press, Columbia.

Hillestad, W., and S. Bennett. 1982. Set-aside areas, National Environmental Research Park. DOE Savannah River National Evironmental Research Park, SRO-NERP-2, Aiken, S.C.

Martof, B. S. 1956. *Amphibians and reptiles of Georgia, a guide.* University of Georgia Press, Athens.

Martof, B. S., W. M. Palmer, J. R. Bailey, and J. R. Harrison III. 1980. *Amphibians and reptiles of the Carolinas and Virginia.* University of North Carolina Press, Chapel Hill.

McFarlane, R. W., R. F. Frietsche, and R. D. Miracle. 1978. *Impingement and entrainment of fishes at the Savannah River Plant.* E. I. duPont de Nemours and Company DP-1491, Aiken, S.C.

———. 1979. Community structure and differential impingement of Savannah River fishes. *Proc. Ann. Conf. Southeastern Assoc. of Fish & Wildlife Agencies* 33:628–638.

Mount, R. H. 1975. *The reptiles and amphibians of Alabama.* Auburn Printing Company, Auburn, Ala.

Myers, C. W. 1967. The pine woods snake, *Rhadinaea flavilata* (Cope). *Bull. Florida State Mus. Biol. Sci.* 11:47–97.

Neill, W. T. 1971. Last of the ruling reptiles: alligators, crocodiles and their kin. Columbia University Press, New York.

Odum, E. P., and E. J. Kuenzler. 1963. Experimental isolation of food chains in an old-field ecosystem with the use of phosphorus-32. In *Radioecology*, Proc. First Natl. Symp. Radioecology, ed. V. Schultz and A. W. Klement, Jr., Reinhold Publishing Corporation, New York.

Schalles, J. F. 1979. Comparative limnology and ecosystem analysis of Carolina bay ponds on the upper Coastal Plain of South Carolina. Ph.D. dissertation, Emory University, Atlanta, Ga.

Schalles, J. F., R. R. Sharitz, J. W. Gibbons, G. J. Leversee, and J. N. Knox. 1989. Carolina bays of the Savannah River Plant, Aiken, South Carolina. DOE Savannah River National Evironmental Research Park, SRO-NERP-2, Aiken, S.C.

Schmidt, K. P., and D. D. Davis. 1941. *Field book of snakes*. J. P. Putnam's Sons, New York.

Sharitz, R. R., and J. W. Gibbons. 1982. *The ecology of southeastern shrub bogs (pocosins) and Carolina bays: A community profile*. U.S. Fish and Wildlife Service, Division of Biological Services, Washington, D.C. FWS/OBS-82/04.

Smith, H. M. 1946. *Handbook of lizards. Lizards of the United States and of Canada*. Comstock Publishing Company, Ithaca, N.Y.

Smith, H. M., and E. D. Brodie, Jr. 1982. *A guide to field identification: Reptiles of North America*. Golden Press, New York.

Tinkle, D. W., and R. Ballinger. 1972. *Sceloporus undulatus*: A study of the intraspecific comparative demography of a lizard. *Ecology* 53:570–584.

Van Pelt, A. F. 1966. Activity and density of old-field ants of the Savannah River Plant, South Carolina. *J. Elisha Mitchell Sci. Soc.* 82:35–43.

Wein, G. R., S. Kroeger, and G. J. Pierce. 1987. Lacustrine vegetation establishment within a cooling reservoir. In *Proc. 14th Ann. Conf. Wetlands Restoration and Creation*, ed. F. J. Webb, pp. 206-218, Hillsborough Community College, Plant City, Florida.

Wright, A. H., and A. A. Wright. 1949. *Handbook of frogs and toads of the United States and Canada*. Comstock Publishing Company, Ithaca, N.Y.

———. 1957. *Handbook of snakes of the United States and Canada*. 2 vol. Comstock Publishing Company, Ithaca, N.Y.

Species Index

Acris
 crepitans, 22, 39, 44, 68, 106
 gryllus, 22, 39, 44, 68, 106
Agkistrodon
 contortrix, 24, 54, 91, 97, 109
 piscivorus, 24, 54, 92, 97, 109
Alligator
 mississippiensis, 23, 75, 98, 107
Ambystoma
 cingulatum, 33, 34, 96
 mabeei, 33, 96
 maculatum, 22, 29, 33, 34, 61, 99, 100, 106
 opacum, 22, 29, 33, 34, 62, 106
 talpoideum, 22, 29, 33, 34, 62, 84, 85, 89, 92, 106
 tigrinum, 22, 29, 33, 34, 63, 64, 106
Amphiuma, 84
 means, 22, 28, 30, 64, 100, 105
Anolis
 carolinensis, 20, 23, 53, 81, 108
Apalone (see *Trionyx spiniferus*)

Bufo
 americanus, 96
 quercicus, 22, 38, 41, 68, 106
 terrestris, 22, 38, 43, 68, 69, 85, 106
 woodhousei, 22, 38, 43, 94, 96, 98

Carphophis
 amoenus, 23, 56, 83, 100, 108
Cemophora
 coccinea, 23, 57, 83, 108

Chelydra
 serpentina, 23, 49, 75, 76, 97, 107
Chrysemys
 picta, 23, 50, 76, 100, 107
 scripta (see *Trachemys scripta*)
Clemmys, 77
 guttata, 23, 50, 76, 108
Cnemidophorus
 sexlineatus, 23, 53, 81, 108
Coluber
 constrictor, 23, 56, 57, 83, 108
Crotalus
 adamanteus, 93, 96, 97
 horridus, 24, 54, 92, 97, 109

Deirochelys
 reticularia, 23, 51, 77, 97, 108
Desmognathus, 28, 32, 64, 94, 96, 98, 106
 auriculatus, 22, 28, 32, 64, 94, 106
 fuscus, 22, 28, 32, 94, 96, 98
Diadophis
 punctatus, 23, 56, 83, 108

Elaphe
 guttata, 23, 59, 84, 108
 obsoleta, 23, 60, 84, 108
Eumeces
 egregius, 96
 fasciatus, 23, 53, 81, 82, 108
 inexpectatus, 23, 52, 81, 82, 108
 laticeps, 23, 53, 81, 82, 108

Species Index

Eurycea
  bislineata (see *Eurycea cirrigera*)
  cirrigera, 22, 25, 28, 37, 64, 106
  guttolineata (see *Eurycea longicauda*)
  longicauda, 22, 25, 28, 37, 65, 106
  longicauda guttolineata, 106
  quadridigitata, 22, 27, 31, 65, 106

Farancia
  abacura, 23, 56, 84, 85, 108
  erytrogramma, 23, 57, 85, 109

Gastrophryne
  carolinensis, 22, 38, 41, 69, 107
Gopherus
  polyphemus, 96

Hemidactylium
  scutatum, 31, 96
Hemidactylus
  turcicus, 98
Heterodon
  platyrhinos, 23, 25, 59, 85, 109
  platirhinos (see *Heterodon platyrhinos*)
  simus, 23, 59, 85, 109
Hyla
  andersoni, 46, 96
  avivoca, 22, 39, 45, 69, 106
  chrysoscelis, 22, 39, 45, 70, 95, 106
  cinerea, 22, 40, 46, 70, 106
  femoralis, 22, 39, 45, 70, 106
  gratiosa, 22, 40, 46, 70, 106
  ocularis (see *Pseudacris ocularis*)
  squirella, 22
  versicolor, 70, 95, 96, 98, 106

Kinosternon
  bauri, 23, 49, 77, 78, 98, 99, 107
  odoratum (see *Sternotherus odoratus*)
  subrubrum, 23, 49, 77, 78, 98, 107

Lampropeltis
  calligaster, 96, 99
  getulus, 23, 59, 86, 109
  triangulum, 24, 59, 86, 99, 109

Limnaeodus
  ocularis (see *Pseudacris ocularis*)

Masticophis
  flagellum, 24, 57, 86, 109
Micrurus
  fulvius, 24, 57, 93, 97, 109

Natrix (see *Nerodia*)
Necturus
  punctatus, 22, 29, 31, 65, 105
Nerodia
  cyclopion, 24, 25, 60, 87, 109
  erythrogaster, 24, 25, 60, 87, 109
  fasciata, 24, 25, 60, 87, 95, 109
  sipedon, 24, 25, 60, 87, 95, 96, 99, 109
  taxispilota, 24, 25, 60, 87, 109
Notophthalmus
  viridescens, 22, 28, 32, 66, 85, 106

Opheodrys
  aestivus, 24, 57, 88, 109
Ophisaurus
  attenuatus, 23, 52, 82, 99, 108
  ventralis, 23, 52, 82, 108

Phrynosoma
  cornutum, 98
Pituophis
  melanoleucus, 24, 59, 88, 109
Plethodon
  glutinosus, 22, 28, 29, 66, 106
  websteri, 66, 96, 99
Pseudacris
  brimleyi, 22, 39, 46, 95, 96, 98
  crucifer, 22, 25, 39, 44, 46, 71, 107
  nigrita, 22, 39, 47, 71, 95, 107
  ocularis, 22, 25, 38, 44, 45, 71, 100, 107
  ornata, 22, 39, 47, 71, 107
  triseriata, 22, 39, 44, 47, 95, 96, 98, 100, 107
Pseudemys
  concinna, 23, 25, 51, 78, 79, 100, 107
  floridana, 23, 25, 51, 79, 97, 107
  scripta (see *Trachemys scripta*)

Pseudobranchus
  striatus, 29, 96
Pseudotriton
  montanus, 22, 28, 37, 67, 106
  ruber, 22, 28, 35, 67, 106

Rana, 85
  areolata, 22, 40, 48, 72, 100, 107
  catesbeiana, 22, 40, 47, 72, 97, 107
  clamitans, 22, 40, 48, 73, 107
  grylio, 23, 40, 48, 73, 97, 99, 107
  heckscheri, 40, 96, 98, 99
  palustris, 23, 40, 47, 73, 99, 100, 107
  sphenocephala, 23, 25, 40, 48, 73, 107
  utricularia (see *Rana sphenocephala*)
  virgatipes, 23, 40, 48, 74, 107
Regina
  rigida, 24, 25, 58, 88, 99, 109
  septemvittata, 24, 58, 88, 100, 109
Rhadinaea
  flavilata, 24, 89, 99, 100, 109

Scaphiopus, 85
  holbrooki, 22, 38, 41, 74, 106
Sceloporus
  undulatus, 23, 53, 81, 82, 108
Scincella
  lateralis, 23, 52, 82, 108
Seminatrix
  pygaea, 24, 56, 89, 109

Siren
  intermedia, 27, 30, 67, 68, 106
  lacertina, 22, 27, 30, 68, 100, 106
Sistrurus
  miliarius, 24, 54, 93, 97, 109
Stereochilus
  marginatus, 35, 96
Sternotherus
  odoratus, 23, 25, 49, 79, 107
Storeria
  dekayi, 24, 58, 90, 109
  occipitomaculata, 24, 57, 90, 109

Tantilla
  coronata, 24, 56, 90, 109
Terrapene
  carolina, 23, 49, 79, 108
Thamnophis
  sauritus, 24, 57, 90, 109
  sirtalis, 24, 57, 91, 99, 109
Trachemys
  scripta, 23, 80, 97
Trionyx
  ferox, 96
  spiniferus, 23, 25, 49, 80, 97, 108

Virginia
  striatula, 24, 58, 91, 100, 109
  valeriae, 24, 56, 91, 99, 109